From Physical Obstacles to Visible Reminders: The Evolution of Traffic Calming

Sanjay

Contents

Chapter 1

Introduction

1.1   Background

The Institute of Transportation Engineers defined traffic calming as "The combination of mainly physical measures that reduce the negative effects of motor vehicle use."; this was done to change driver behaviour, thus improving safety for other street users [1].

Many speed calming techniques require physically altering the road to prohibit a vehicle from moving at restricted speeds, and usually involve making either asphalt humps or smaller evenly spaced asphalt speed bumps called rumble strips. Other less permanent techniques involve placing rubber speed bumps and cones on the road [2].

These physical alterations are effective in preventing drivers from approaching at high speeds. However, they have a negative effect on a vehicle's suspension system when the vehicle drives over them at high speed. Physical alterations on the road are damaging, even when the vehicles are not driving at an exceptionally high speed. These measures accelerate a vehicle's suspension system's wear and tear, making them unpopular in many areas [2]. Table A.1-1 in Appendix A.1 gives a detailed account of the various traffic calming techniques and their effectiveness in reducing vehicle speeds and traffic volumes.

Speed signs are the most common method of speed calming in South Africa. They are boards on either side of the road with an illustration of a particular speed limit for a given strip of road. In South Africa, speed signs typically start from 20 km/h up to 120 km/h [3]. Traffic officers routinely monitor them with Doppler radars that have cameras; this is done to ensure compliance from the drivers. Unfortunately, compliance is temporary since there are limited traffic officers on the road and cannot enforce compliance at every speed sign [3]. Thus, there was a need for a more visible and automated form of speed calming.

Radar speed signs allow for a visible and automated form of speed calming since the radar measures the vehicle's radial speed and displays it to the driver in real-time without the need of a dedicated operator.

Chang et al. [4] had made a scaled study into radar speed signs' effectiveness as traffic calming devices in neighbourhood streets. This study employed four radar speed signs placed strategically across King County, Washington, along 108th Avenue NE between NE 124th Street and Juanita-Woodinville Way NE, as shown in Figure 1-1. To evaluate the effectiveness of these signs, speed measurements were conducted before, during, and after the installation. The authors then concluded that radar speed signs did represent a form of traffic calming and that these signs have shown to be useful devices with sustained traffic safety benefits [4].

Figure 1-1: Radar speed sign locations on representative map [4]

1.2   Problem Statement and Research Motivation

The purpose of this study was to design a traffic calming strategy for the CSIR campus in Pretoria. This traffic calming strategy intends to reduce vehicle speeds, improve transit access, and protect pedestrians and animals on campus.

The CSIR currently employs a combination of traffic calming strategies, including traffic circles, speed limits, speed humps, and radar-based speed signs, to protect pedestrians and animals on campus. The radar-based speed sign system employed at the CSIR campus was provided by called Cool-Ideas (Pty) Ltd. This company imports these systems from a company named Houston Radars Inc., based in the United States of America (USA), at a relatively high cost, which inflates the prices in which the CSIR pays for the devices.

Other companies that provide these systems in South Africa (SA) include Truvelo, Repro supplies, Polycomp, and all have confirmed telephonically that they import their radars from either Houston radars, Wanco Inc., Trafficlogix, Monitor systems, Radar sign, also known as driver feedback signs or MPH Industries and Photon play systems. These American companies only design the housing and physical structure of the radars and implement the radar signal processing for these systems, but they do not manufacture the actual radar sensors used in these systems.

South African companies import these systems at a very high cost, which factors in the fluctuation of both the United States Dollar (USD) and South African Rand (ZAR) as well as company profit margins. These factors cause these systems to be costly, each unit typically costing around R60 000 excl[1]. VAT in SA. These factors culminated in a collaboration between Cool-Ideas (Pty) Ltd and the CSIR to investigate the development of a low-cost traffic calming radar.

This type of radar would enable the management of office parks and business campuses like the CSIR to be able to monitor and regulate the vehicle speeds on their properties. Radars that are used for this particular application are compact devices that house the radar sensor, signal processing block, and power supply system. They have a speed display that acts as a feedback mechanism.

The speed display allows the driver to see their speed, and if they are travelling at higher speeds than permitted, they can apply corrective measures immediately. An example of a speed calming radar was shown in Figure 1-2.

---

[1] Based on a quotation supplied by Cool-Ideas (Pty) Ltd.

Figure 1-2: Radar speed sign. Components include speed sign, speed display and radar module. The system was powered by a solar panel [5].

1.3   User Requirements

This study aimed to formulate a detailed "blueprint" of a traffic calming radar for the use on the CSIR business campus to ensure the safety of non-motorized road users and the animals. The following requirements outline functional specifications the radar-based traffic calming solution needs to fulfill to be deemed suitable for this application. The following requirements were developed together with stakeholders from the CSIR.

1. The system shall detect small vehicles at a distance of 40 m.

2. The system shall have velocity estimation accuracy comparable to current commercial and experimental systems.

3. The system shall provide speed measurements that must be visible to vehicles from 40m away.

4. The system shall be robust and have components rated for the use of up to two years.

5. The system shall be operational in the day as well as the night.

6. The total system components and labour shall not cost more than R20k with components costing less than R15k.

1.4   Research Gap

Many researchers and institutions are interested in the vehicle velocity measurement problem under urban traffic and access-controlled campuses. Technical solutions are available in the market [6] and Litman et al. [7] also give a detailed account of the different speed calming techniques commonly used to ensure the safety of non-motorized road users. Unfortunately, these reports do not detail the specifications to enable interested parties to develop and implement low-cost radar-based speed signs. Thus, there is a need for documentation that follows a system engineering approach and outlines such devices' design and implementation.

This study gives a procedural approach to designing a low-cost continuous wave (CW) Doppler radar for traffic calming. Riid et al. note the sparsity in the scientific literature of using low-cost microwave radar for traffic monitoring [8], hence this study also contributes to using this technology for this particular use case. The performance of the proposed system was benchmarked against the given specifications of established commercial systems to ensure industry-related performance.

The Key Performance Indicators (KPI) used to gauge the performance of the radar-based traffic calming solution includes the radars performance, signal processing latency, velocity estimate visibility as well as reliability of the system are described in Table 1-1.

Table 1-1: Key performance indicators for the radar

Key Performance Indicators (KPI)	Description
Detection rate	The measure of how precisely the radar system correctly identifies a target given a set of measurements.
Maximum detection range	The largest distance the radar can positively identify an object as a target.
Velocity measurement accuracy	The margin of error attributed to the velocity estimates made by the radar.
Signal processing latency	The duration taken by the radar to process and display the velocity estimate.
Visibility and readability of the velocity estimate	Visibility and readability of the velocity estimate from the maximum target detection distance.
Reliability of the system	Reliability of the system to operate as intended for a period comparable to that of established commercial systems; this includes having power redundancies to ensure the day and night-time operation.
Cost	The cost associated with assembling the system, including labour and hardware components.

The development of the signal processing software was not accounted for in the system's total cost; this entails counting hours spent researching, developing, and testing the software at the hourly rate of the responsible engineer. Taking this cost would significantly inflate the cost of the individual unit price of the proposed system. Since this was a one-off cost, it does not have a significant bearing in the overall cost of systems produced after that.

1.5 Initial Investigation

The initial investigation of this project described the problem context and aimed to answer the following questions:

1. What is the state of the art experimental and commercial radar speed signs in terms of their technical capabilities and specifications?
2. What are the advantages and disadvantages of these systems?
3. What was the cost associated with purchasing the systems?

1.6 Problems to be Investigated

The study also investigated the following problems in this project:

1. What are the technical considerations that were accounted for when designing a radar-based traffic calming system that avoided misdetections of cars?

1. What is the precision and accuracy of the radar-based traffic calming system for different velocity measurements when measuring a single vehicle travelling towards the system?
2. How much does it cost to build a radar-based traffic calming radar that is comparable to the current commercial and experimental system in performance?

1.7 Project Objectives

This study seeks to investigate and analyse specific traffic calming techniques that will enable the campus management to enforce set speed limits for vehicles travelling inside access-controlled campus roads. Most benefits, costs, and impacts of traffic calming techniques have been studied and detailed in the report by Litman et al. and a high-level summary was illustrated in Table A.1-1 [7]; while Rajani et al. [2] details the benefits, design, and implementations of traditional and smart speed bumps.

Thus, the project objective was to develop an integrated radar system intended to be used as a radar speed sign for traffic calming purposes, since their effectiveness was established by Chang et al. [4].

This objective was further divided into the following actionable tasks:
- Clearly state the technical requirements of a radar-based traffic calming system in line with user requirement specifications.
- Design, construct and test a radar prototype that precisely and accurately estimates the velocity of a single vehicle approaching the radar within reasonable margins of error.
- Quantify the radar-based traffic calming system's performance using recorded data against key performance indicators such as radar performance, signal processing latency, velocity estimate visibility, and reliability of the system.
- Ensure that the proposed system's total cost does not exceed the user budget while providing key performance targets comparable to commercial systems.

1.7.1 Project Scope

This project was carried out under specific conditions; this was done to limit the scope of work to fit the project's limited time frame and budget.

- This project consisted of building a radar-based traffic calming technology demonstrator.
- The project was carried out in 1 year, 3 months.
- All experiments were carried out at the CSIR Pretoria campus, where the technology demonstrator was either placed on the side of the dry asphalt road or directly on the dry asphalt road. The scenes had vegetative clutter as well as returns from the main building only. No experiment was done next to a metal structure or directly facing a metal structure.
- The BMW i3 and Toyota Yaris were referred to as the targets of interest in this study.
- All experiments were carried out in clear weather conditions, such as typical Pretoria summer conditions with temperatures ranging from 25° C to 32° C with no wind, rain, nor hail.
- The vehicles' typical orientation was front-facing as the car approaches the radar and rear-facing as they moved away from the technology demonstrator.
- The measurements were limited to one vehicle at a time.
- The technology demonstrator's components did not cost more than R5 000 to purchase, which was the project's budget.
- The developed technology demonstrator results were intended to be compared to other k-band industry-standard radars and were not to be compared with radars developed at the CSIR.
- Only the developed system was used to measure vehicles, and industry-standard systems were only compared using the specifications detailed in their datasheets.

1.7.2 Exclusions

The project objective did not require the following:

- To design and develop dedicated hardware for the radar module.
- To procure all mounts and associated hardware to have a fully autonomous system that is ready for deployment and commercialization.
- Assemble any recommendations made by the system design.

The technology demonstrator constructed is for the sole purpose of demonstrating feasibility for a traffic calming system, and the insights gained could form the specification inputs for product prototype development.

1.8 Dissertation Overview

This document is divided according to the following chapters, each detailing the process taken to fulfill the user requirements.

1.8.1   Chapter 2: Literature Study

An in-depth review of current experimental and commercial systems was explored to understand the technological landscape. This review helped to formulate the user requirements that shaped the contents of this design project.

1.8.2   Chapter 3: Theoretical Considerations for Radar Modelling

Chapter 3 focuses on primary radar and signal processing fundamentals to understand the underlying principles that govern radar systems in general and CW Doppler radar in particular; this would enable effective system design and testing.

1.8.3   Chapter 4: System Requirement Development and Methodology

This chapter outlined the technical and non-technical requirements in which this project was measured. The method executed for completion of this project was outlined. The experiments that determined the reliability and accuracy of this system were also stated.

1.8.4   Chapter 5: System Design and Simulations

This chapter outlined the design by using system requirements to obtain the design parameters. The radar module, ADC, digital signal processer (DSP), display module, and the power system were all specified and selected in this chapter.

1.8.5   Chapter 6: System Integration and Testing

This chapter outlined the system integration and the performance benchmark of each sub-system. The issues that arose during the integration processes were outlined and addressed.

1.8.6   Chapter 7: Acceptance Testing and Results

The integrated data-collection system was tested against each technical requirement to assess if the system's technical requirements were met. Then an in-depth analysis of the results was made detailing the performance of the system with each KPI.

1.8.7   Chapter 8: Conclusion and Future Work

This section outlined which of the requirements were satisfied and which were not. Recommendations were also stated on how to improve the design further and also give insight into future work.

Chapter 2

Literature Review

2.1 Radar Development

Radar (Radio Detection and Ranging) systems use electromagnetic (EM) waves for various applications such as imaging stationary and moving objects. Analyzing the reflected EM wave characteristics can enable detection and classification objects of interest and extract other characteristics such as object state.

Radars today have been developed to come in many configurations that suit different applications. These applications extend well beyond the military domain in which radar was initially introduced. Radars are used in several consumer electronic devices, industrial applications, law enforcement, and in the automotive industry [9].

2.2 Experimental and Commercial Radar Speed Signs

The following was an in-depth review of the current radar speed sign landscape for commercial and experimental use cases. These radars' specifications formed part of the performance baseline standard for this dissertation's radar system design.

2.2.1 Experimental Radar Speed Detectors

Hobbyists, researchers, and radar enthusiasts developed the experimental radar speed detectors featured in this section. These radars are not intended for a commercial setting or law enforcement but were developed as technology demonstrators.

Radar speed detector based on Raspberry Pi

The first radar speed detector investigated is a Raspberry Pi-based CW speed detector tasked with recording traffic data detailed by Butterfield et al. [11]. This system was not designed for law enforcement but rather to record the rate of traffic flow on a stretch of road. This system creates a time series bar chart with data about the traffic flow in the region of interest for a certain extent of time [10].

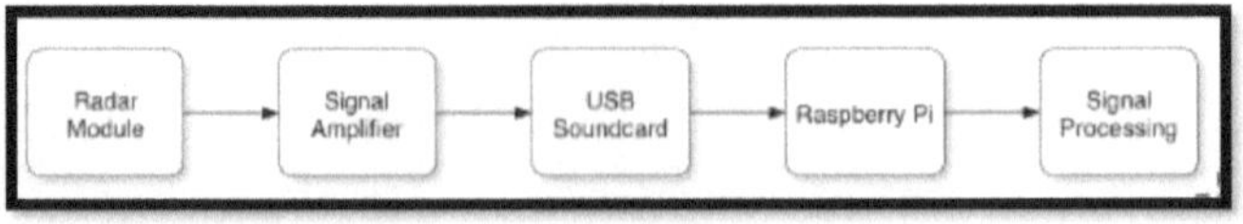

Figure 2-1: High level experimental radar speed detector system architecture [10].

Figure 2-1 illustrates that the system consists of five main sub-systems. The first is a radar module that captures target echoes from transmitted signals; the signal amplifier increases the measured echoes' amplitude. The next sub-system was a USB soundcard with a 16-bit analogue-to-digital converter (ADC) with a signal to quantization noise ratio (SQNR) of 186.62 dB, since most of the signals that are captured are mixed down into the audible range by the radar module [11]The last two sub-systems are contained within the Raspberry Pi and consist of software to record the data and run the signal processing script [10].

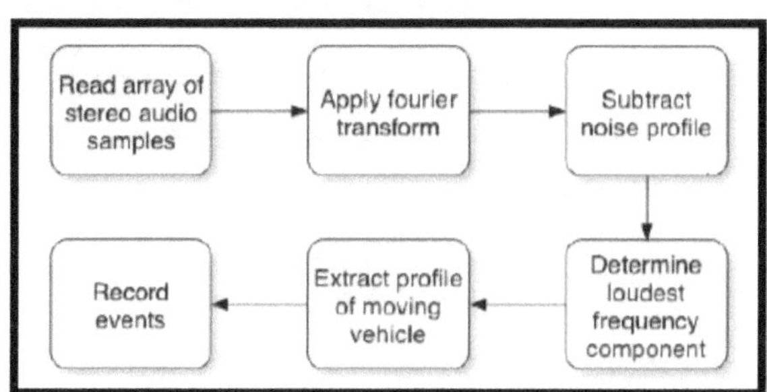

Figure 2-2: Signal processing block diagram [10].

Figure 2-2 illustrates the signal processing chain, which includes reading the I/Q samples from the ADC (sampling), then applying a Fourier transform to obtain the signal's frequency components (Doppler processing). The next step in the processing chain was the noise profile subtraction, where the frequency components of the noise are digitally filtered. Then a threshold detector was used to extract the frequency with the highest amplitude from the signal, which was a detection and the Doppler frequency (radial velocity) measurement. This processing chain's last step was to save the data into a CSV file for further analysis/documentation.

The hardware used includes an RSM2650 24 GHz Stereo radar sensor. This module has an operating voltage of 4.75 V and a supply current of 30-40 mA. The sensor can operate between -20°C and 60°C and produce a maximum output power of 16 dBm. The antenna characteristics include an 80° azimuth (Az) beam-width and a 32° elevation (El) beam-width; there was no mention of the antenna gain in the datasheet. The maximum sidelobe levels were -16 dB in Az and -21 dB in El. The physical sensor dimensions are 25 x 25 x 12.7 mm.

The signal amplifier used was an LM833n low noise operational amplifier with a power bandwidth of 120 kHz—a 5 V DC supply powers this sub-system. The amplifier has a dynamic range (DR) of about 140 dB. Figure 2-3 shows the collected data over four days. The data saved includes incoming cars and outgoing cars. The full system specifications were detailed in Table 2-1.

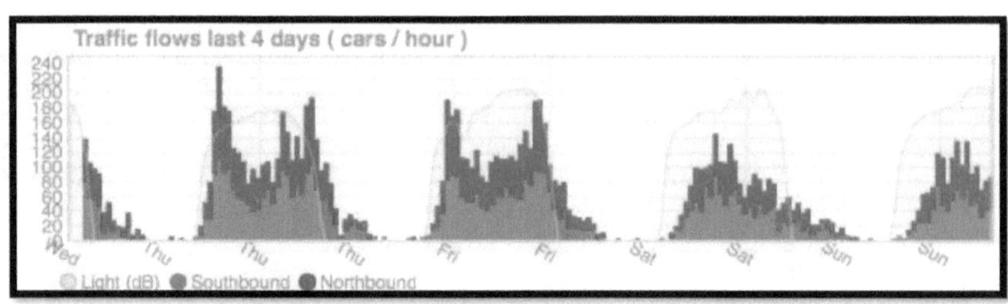

Figure 2-3: Time series bar chart with 4 days of traffic flow [10].

Table 2-1: Raspberry Pi based CW speed detector's system specifications [10].

Parameters	Specifications	Units
Supply voltage	5	V
Carrier frequency (F_c)	24	GHz
Max Doppler shift	11.03	kHz
DR	140	dB
SQNR	186.62	dB
Operating current	30-40	mA
Power output (EIRP)	16	dBm
Antenna beam-width (Az x El)	80 x 32	°
Sidelobe levels (Az x El)	-16 x -21	dB
Temperature rating	20 to 60	°C
Physical dimensions (L x W x H)	25 x 25 x 12.7	mm

This system has the following shortcomings making it unsuitable for a traffic calming application. The most prominent being that this system does not display the speed readings in real-time. The system also does not have a false alarm detector; this means that any signal that was within the bandwidth of the radar module with a sufficient amplitude would be captured as a target. The system does not distinguish between two competing signals; consequently, it cannot be known with certainty where each signal originates. For instance, if two cars are within the beam and both are travelling at the same speed, and in the same direction, the system treats the resultant signal as the same, single target.

STM32L476 based Radar

The next system studied is a radar speed detector based on the STM32L476 discovery board by Longstaff-Tyrrell et al. [12]. Figure 2-4 illustrates the high-level design of the STM32L476 based radar speed detector, which includes a radar module, an STM32L476 Discovery board, and a speed display.

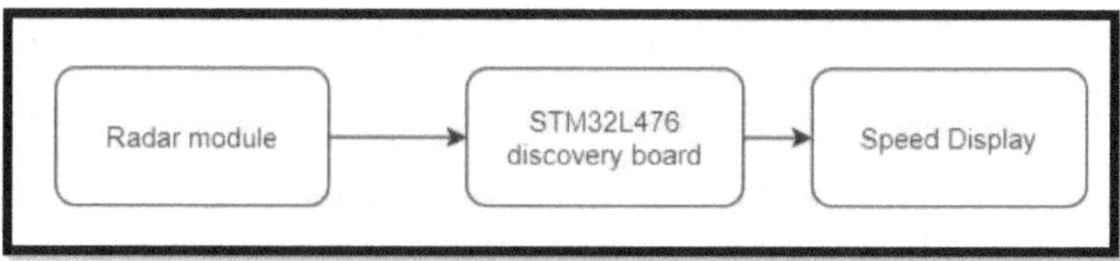

Figure 2-4: High level design of STM32L476 based radar speed detector.

The radar module is the HB100 an x-band 10GHz CW motion sensor that has an Equivalent Isotropically Radiated Power (EIRP) of 20 dBm and spurious emission of -7.3 dBm. The sensor has an operating voltage of 5V DC and maximum current consumption of 40 mA. The antenna beam-width at -3 dB in azimuth and elevation was 80° and 40° respectively. The sensor operating temperature was between -15°C and 55°C.

Figure 2-5 illustrates the sub-systems contained by the STM32L476 discovery board.

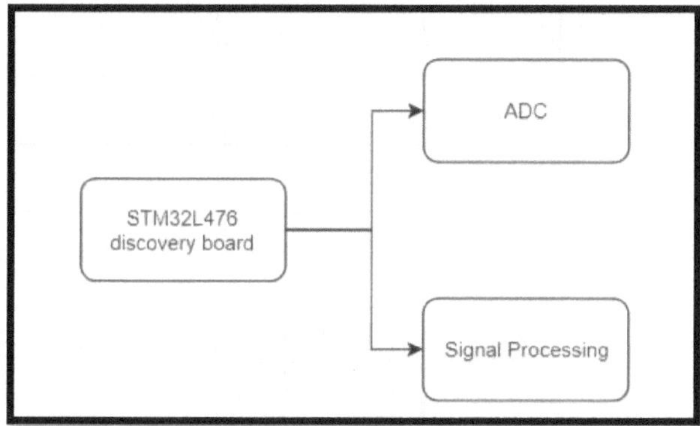

Figure 2-5: STM32L476 sub-systems.

The analogue to digital converter has a 12-bit resolution; this, according to [12], gives the system an SQNR of 73.88 dB, which results in relatively low distortion in the magnitude spectra. This system applies a fast Fourier transform (FFT) to obtain the radar output spectrum. The system achieves an accuracy of ±1 km/h; the FFT length was 512 samples, which, at a sampling rate of 8 kHz, produces a frequency resolution of 15 Hz and a DR of 66.22 dB. This system was powered by a 5 V DC input. The system specifications were detailed in Table 2-2. The major advantages of this system are its ability to provide near real-time velocity estimate display as well as its fine frequency resolution. The disadvantage of this system was that it did not have a false alarm detector.

Table 2-2: STM32L476 discovery board's full system specifications [12].

Parameters	Specifications	Units
Supply voltage	4.75 and 5.25	V
Carrier frequency (F_c)	10.525	GHz
Max Doppler shift	4	kHz
DR	66.22	dB
SQNR	73.88	dB
Operating current	40	mA
Power output (EIRP)	20	dBm
Antenna beam-width (Az x El)	80 x 40	°
Sidelobe levels (Az x El)	-18 x -12	dB
Temperature rating	-15 to 55	°C
Physical dimensions (L x W x H)	25 x 25 x 12.7	mm

Low-cost Vehicle Detection and Classification System based on Unmodulated Continuous-wave (LDC-CW)

The next system studied is a low-cost vehicle detection and classification system based on unmodulated CW radar (LDC-CW) proposed by Fang et al. [13]. This system was used in an intelligent transportation system. The radar utilizes a k-band unmodulated CW radar module with a carrier frequency of 24.125 GHz, it features time-frequency analysis, multi-threshold detection, and Hough transforms as the major signal processing methods to extract the speed of the vehicles from the Doppler signatures generated by the targets.

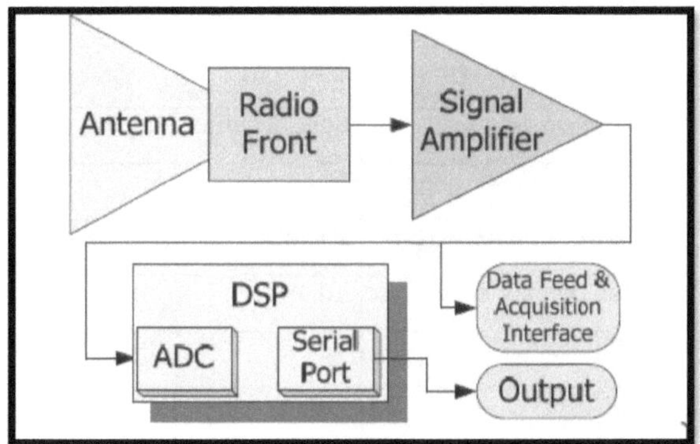

Figure 2-6: Block diagram of the vehicle and classification system [13].

Figure 2-6 shows that the system consists of 3 sub-systems: the radio frequency front end (RFFE), the signal amplifier, and the DSP. The signal processing framework of this device is illustrated in Figure 2-7 [13].

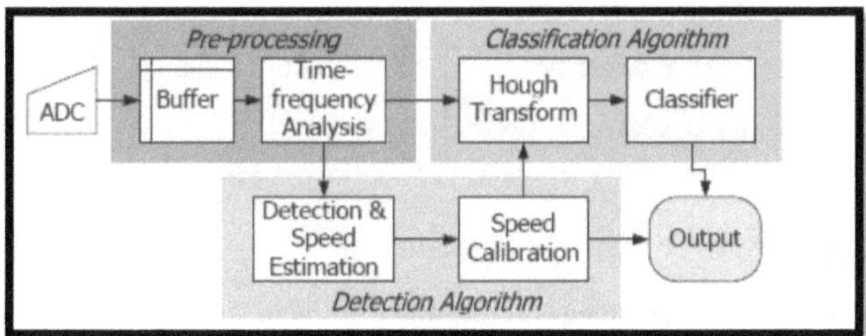

Figure 2-7: Signal processing framework [13].

This system has an antenna gain of 17 dBi, the gain of the analogue amplifier was 60 dB, the amplifier bandwidth was 10 kHz, and the sampling frequency was 20 kHz. The output power of the radar was 5 mW, and the receiver sensitivity was -90 dBc. This system was based on a Texas Instruments TMS320F2808 DSP with a 12-bit ADC and a 32-bit central processing unit (CPU). The 12-bits found in the ADC give the system an SQNR of 73.88 dB and a DR of 66.22 dB.

The physical dimensions of the system are 4 cm x 5 cm x 16 cm. The accuracy of the speed measurement was 97.1% or $\pm$ 1 km/h, and the maximum speed that can be captured was 225 km/h. The system was described as having a very narrow beam-width; thus, only one car can be detected at a time.

The system had an average detection rate of above 9%, and the average accuracy of the speed measurements was 97.1%. The classification algorithm had an average performance of 94.8% across all studied vehicle types. LDC-CW's system parameters were described in Table 2-3 and the total cost of the system was $200 USD [13].

Table 2-3: LDC-CW's system specifications [13].

Parameters	Specifications	Units
Supply voltage	5	V
Carrier frequency (F_c)	24.125	GHz
Max Doppler shift	10.05	kHz
DR	66.22	dB
SQNR	73.88	dB
Operating current	40	mA
Power output (EIRP)	6.99	dBm
Antenna beam-width (Az x El)	20 x 30	°
Sidelobe levels (Az x El)	N/A	dB
Temperature rating	-15 to 55	°C
Physical dimensions (L x W x H)	40 x 50 x 160	mm

Universal Radar (uRAD)

The next experimental system studied was the universal radar (uRAD) developed by Anteral S.L. [14]; this radar was in the form of a shield for the use with a Raspberry Pi or Arduino. The system's physical dimensions are 18 mm x 67.5 mm x 53.5 mm and weigh 13 g. The system was shown in Figure 2-8.

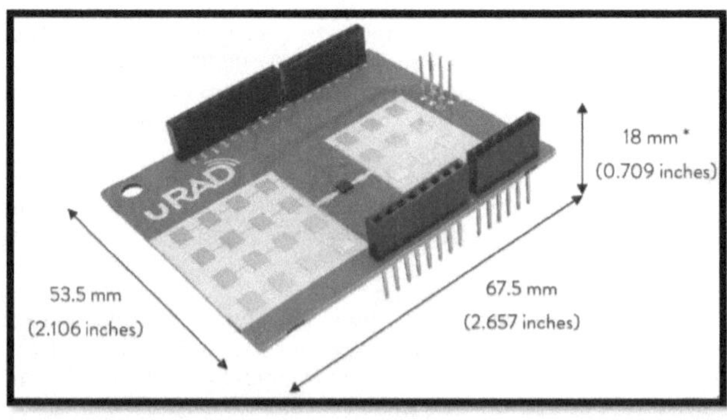

Figure 2-8: uRAD experimental radar shield Arduino version [14].

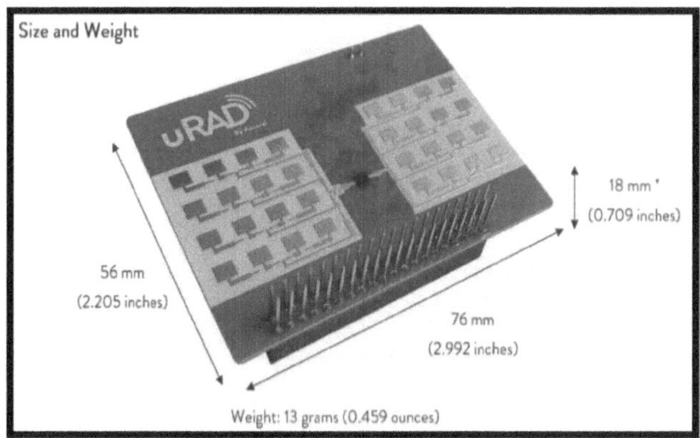

Figure 2-9: uRAD experimental radar shield for Raspberry Pi [14].

The system utilizes an array of patch antennas with a field of view of 20° and 18° in the elevation and azimuth, respectively, at -3 dB beam-width for the Tx antenna. The Rx antenna has a beam-width at -3 dB of 30° and 21° in elevation and azimuth. The system has 2 modes, CW and FMCW. The maximum output power was 20 dBm.

The sidelobe level of the RX antenna was -14.8 dB and -13.4 dB in elevation and azimuth, respectively. In CW mode, the system has a velocity range from 2.52 km/h to 270 km/h and a velocity accuracy of ±0.18 km/h. The maximum detection distance of a car with an RCS of 12.5 dBsm is 75 m.

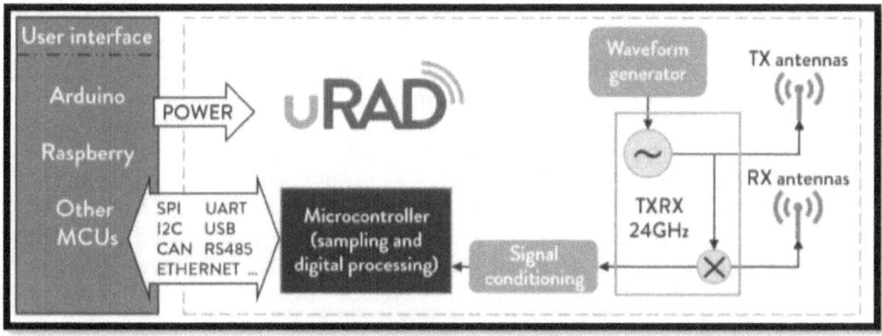

Figure 2-10: uRAD system overview [14].

he system has four main sub-systems, the radar module, signal conditioner, microcontroller, and the user interface, as shown in Figure 2-10, he system specifications are summarized in Table 2-4. The total cost of the system was €199. The total cost of the system was €199. The system's advantages include its relatively long detection range and built-in signal processing algorithms, such as a false alarm rate detector. The main disadvantage was the system's relatively high cost [14].

Table 2-4: uRAD experimental radar shield's system specifications [14].

Parameters	Specifications	Units

Supply voltage	3.5 to 10	V
Carrier frequency (F_c)	24.125	GHz
Max Doppler shift	12.1	kHz
DR	N/A	dB
SQNR	N/A	dB
Operating current	170	mA
Power output (EIRP)	20	dBm
Antenna beam-width (Az x El)	21 x 30	°
Sidelobe levels (Az x El)	-13.4 x -14.8	dB
Temperature rating	-15 to 55	°C
Physical dimensions (L x W x H)	67.5 x 53.5 x 18 76 x 56 x 18	mm

Experimental System 1 (EXP 1)

The next experimental system that was studied was a low-cost microwave radar used to detect vehicles and their speeds and the direction of arrival estimation at the observation point. This study was presented at the Biennial Baltic Electronics Conference (BEC) in 2018 by Riid et al. [8]. This system shall be referred to as experimental system 1 or EXP 1 for short. EXP 1 consists of 4 sub-systems, including an x-band MDU1740 Doppler motion detector by Microwave Solutions, an amplifier module, a DSP, and a detection algorithm.

The motion detector has a transmission frequency of 9.35 GHz and has an operating voltage of 3.6 V $\pm$0.2 V and an operating current of 60 mA. The output power of the sensor was 10 dBm EIRP, and the antenna gain was 8 dBi. Both the Tx and Rx antennae have a field of view of 72° and 36° in the vertical and horizontal planes. The sensor can operate at a temperature of between -10°C to +55°C. The researchers designed a two-stage amplifier that used a precision JFET amplifier with the model number ADA4610-2. The system works in CW mode but is also capable of functioning in pulsed mode. The amplifier circuit contains two high pass filters and two low pass filters, yielding the frequency bounds of 3.4 Hz to 2800 Hz, respectively. The latter being the maximum Doppler frequency that the system can allow therefore limiting the maximum speed measurement to 160 km/h [8]. EXP 1's system specifications are summarized in Table 2-5.

Table 2-5: EXP 1's system specifications [8].

Parameters	Specifications	Units
Supply voltage	3.6 $\pm$0.2	V
Carrier frequency (F_c)	9.35	GHz
Max Doppler shift	1.5	kHz

DR	N/A	dB
SQNR	N/A	dB
Operating current	60	mA
Power output (EIRP)	10	dBm
Antenna beam-width (Az x El)	36 x 72	°
Sidelobe levels (Az x El)	-13.4 x -14.8	dB
Temperature rating	-10 to 55	°C
Physical dimensions (L x W x H)	46 x 10 x 37.7	mm

The specifications of the DSP were not provided in this paper as the data was processed offsite. The system's sampling frequency was 3 kHz; this sampling rate further limited the maximum detectable speed to 86 km/h by the Nyquist criterion. The system achieved a nearly 95% detection rate for cars 3 m away; the velocity estimation performance was not explicitly quantified [8]. An advantage of the system was its ability to detect the car's direction-of-arrival (DoA). A limitation of the system's performance was its inability to distinguish between two or more vehicles that arrive at the sensor location simultaneously.

Experimental System 2 (EXP 2)

The last experimental system that was investigated was traffic monitoring system implemented using standard discrete component technology. This system was detailed in a paper presented in the 4th European Radar Conference by Alimenti et al. [15]. This system shall be referred to as Experimental System 2 or EXP 2. EXP 2 consists of 3 sub-systems, which includes a 24 GHz radar CW sensor, a low noise Hetro-Junction Field Effect Transistor (H-JFET), as well an 8051 microcontroller. EXP 2 has a current consumption of 100 mA at 12 V supply. The radar sensor is composed of a 10 x 4 patch array antenna elements, with an antenna gain of 13 dB. The -3 dB beam-width of the antenna was a $\pm4.5°$ in both azimuth and elevation. EXP 2's system specifications were summarized in Table 2-6.

Table 2-6: EXP 2's system specifications [15].

Parameters	Specifications	Units
Supply voltage	12	V
Carrier frequency (F_c)	24	GHz
Max Doppler shift	3.57	kHz
DR	N/A	dB
SQNR	N/A	dB
Operating current	60	mA

Power output (EIRP)	10	dBm
Antenna beam-width (Az x El)	4.5 x 4.5	°
Sidelobe levels (Az x El)	N/A	dB
Temperature rating	N/A	°C
Physical dimensions (L x W x H)	N/A	mm

The total power output of the sensor was 6 dBm. The low-noise amplifier has a gain of 10 dB. This radar's main advantage was that it had a detection range of 350 m but only for the velocity measurement. The minimum allowable velocity accuracy was calculated to be to ±4 km/h when the vehicle is further than 15 m from the radar travelling at 80 km/h [15]. Unfortunately, the exact results of the accuracy of the velocity measurements were not published and the detection rate of the device. Due to its narrow beam-width, the system can only illuminate one lane at a time, which is an advantage for speed calming applications.

Summary

Table 2-7 summarizes the different experimental radars that were investigated in this section.

Table 2-7: Summary of experimental radar speed detectors

Parameter	Product name					
	Radar speed detector based on Raspberry pi	STM32L476 based Radar	LDC-CW	uRad	EXP1	EXP2
Mode	CW	CW	CW	CW/FMCW	CW/Pulsed	CW
No. of Bits	16	32	12	N/A	N/A	N/A
Sensor Frequency[2]	24 GHz	10 GHz	24 GHz	24 GHz	9.35 GHz	24 GHz
Detection rate	N/A	N/A	95%	90%	95%	N/A
Velocity Estimation Accuracy	± 1 km/h	± 1 km/h	± 2.8 km/h	± 0.18 km/h	N/A	± 4 hm/h
DR	186.62 dB	192.38 dB	68.38 dB	N/A	N/A	N/A
SQNR	194.4 dB	200.14 dB	76.14 dB	N/A	N/A	N/A
DC Power	5 V	5 V	3.3 V	5 V	3.5 V	12 V
Two lane system	Yes	No	Yes	Yes	Yes	No

---

[2] IEEE definition of K-band is a frequency band from 18 to 27 GHz

Max speed	247.5 km/h	180 km/h	225 km/h	270 km/h	86 km/h	~ 80 km/h
Max Detection Range	50 m	25 m	15 m	75 m	~6-7 m	350 m
Weight	1.3 kg	230 g	200 g	500 g	N/A	N/A
Cost[3]	R1441.87	R478.80	R3267,13	R5802,27	R806.98[4]	N/A

The systems discussed above came from various sources, but the topic of low-cost radar sensors being used for traffic monitoring is mostly undocumented in reputable scientific sources. The missing information from sources that were found could be attributed to the proprietary nature of the systems developed [15]- [14].

Topics that are mainly covered include using high-resolution LFMCW radar for Traffic road monitoring [16], traffic surveillance, and road lane detection using radar interferometry [17]. While these topics gave insight into the state-of-the-art algorithms and hardware used in road monitoring and traffic surveillance, these systems use sophisticated scientific equipment used in laboratories, which are neither compact nor economical.

2.2.2   Commercial Radar Speed Signs

The commercial radars reviewed in this section include the speed signs typically procured by South African companies such as Cool-ideas, Truvelo, etc. These systems are found in business campuses, security estates, and upmarket neighbourhoods around South Africa.

PNL10 by Huston Radar

The first commercial system to be studied was the PNL10 radar system developed by Huston Radar [18]. This system consists of a low power radar module and display block. The radar module was a CW module that operates at a center frequency of 24.125 GHz or 24.2 GHz. The maximum radar power output was 6.99 dBm.

The beam-width of the radar module at -3 dB was 38° in Az and 45° in El. The maximum target return range of the radar was 90 m for targets with an RCS of 12.5dBsm. The system's physical dimensions in terms of L x W x H are 406.4 x 28 x 279.4 in mm. These dimensions are illustrated in Figure 2-11. The display unit is a seven-segment LED display with a maximum brightness of 34557.5 lux.

---

[3] The prices shown are converted using the following exchange rates: R16.34/$, R 19,34/€ and R 21,1/£
[4] Cost excluding ADC and DSP.

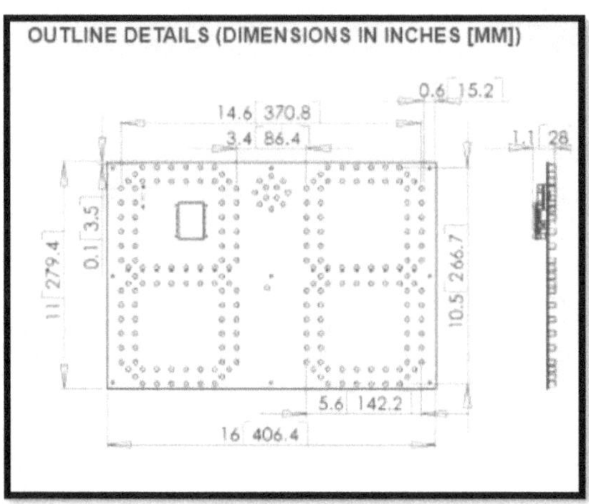

Figure 2-11: Outline details of the PNL10 radar system [18].

This system requires 7.47 W of input power at 18 V DC and 415 mA. The whole structure weighs 450 g and has an operating temperature between -40°C to 85°C [18]. The system diagram was shown in Figure 2-12.

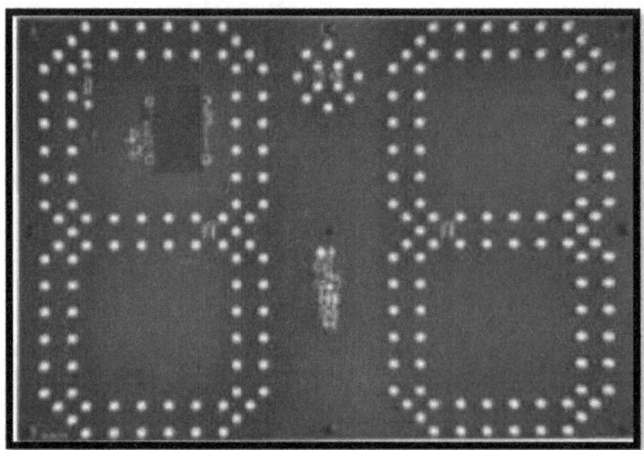

Figure 2-12: PNL10 radar system developed by Huston Radar [18].

The PNL10's radar system specifications were summarized in Table 2-8.

Table 2-8: PNL10's system specifications [18].

Parameters	Specifications	Units
Supply voltage	18	V
Carrier frequency (F_c)	24.125 or 24.2	GHz
Max Doppler shift	10.78	kHz

DR	N/A	dB
SQNR	N/A	dB
Operating current	450	mA
Power output (EIRP)	6.99	dBm
Antenna beam-width (Az x El)	38 x 45	°
Sidelobe levels (Az x El)	N/A	dB
Temperature rating	-40 to 85	°C
Physical dimensions (L x W x H)	406.4 x 28 x 279.4	mm

The system has a velocity measurement accuracy of ± 1.6 km/h, and the maximum detectible speed is 159 km/h. The manufacturer guarantees two years of maintenance-free operation and does not include batteries and solar power options. The system can only detect one vehicle at a time.

Pole-mount RSS developed by Wanco Inc

The second commercial system that was reviewed is called the Pole-mount RSS, which is developed by Wanco Inc [19], shown in Figure 2-13; this is a pole-mounted system that operates in Doppler mode only.

Figure 2-13: Pole-mount RSS developed by Wanco Inc [19].

The Pole-mount RSS consists of three sub-systems: a radar module, a solar panel, and a display unit. The radar module is a k-band CW system operating at 24.125 GHz with a maximum power output of 20 dBm. This sensor can obtain returns at a maximum distance of 305 m. The minimum and maximum detectible target speeds are 8 km/h and 222 km/h. The accuracy of the speed measurements is ± 1.6 km/h for speeds of between 8 km/h to 64 km/h and ± 3.2 km/h for rates greater than 64 km/h up until 161km/h. The accuracy for speeds that are greater than 161 km/h is unspecified. The radar can operate at temperatures between -40°C and 85°C [19].

The solar system has a rated output power of 65 W, providing a voltage of 16.9 V DC at a current of 2.34 A. The system also has a battery system with two group 24 deep cycle batteries. These batteries can provide a 12 V DC each, at a maximum current of 750 mA. Their capacity is rated at 150 Ah at 12V DC. The Wanco Pole-mount RSS' system specifications were summarized in Table 2-9.

Table 2-9: The Wanco Pole-mount RSS' system specifications [19].

Parameters	Specifications	Units
Supply voltage	12	V
Carrier frequency (F_c)	24.125	GHz
Max Doppler shift	9.92	kHz
DR	N/A	dB
SQNR	N/A	dB
Operating current	750	mA
Power output (EIRP)	20	dBm
Antenna beam-width (Az x El)	38 x 45	°
Sidelobe levels (Az x El)	N/A	dB
Temperature rating	-40 to 85	°C
Physical dimensions (L x W x H)	770 x 12 x 640	mm

The display unit consists of an AlInGaP II LED system that was amber in colour. The LED matrix consists of 12 x 10 pixels achieving a maximum brightness of 65973 lux. The system dimensions in L x W x H are 770 x 12 x 640 mm, as shown in Figure 2-14.

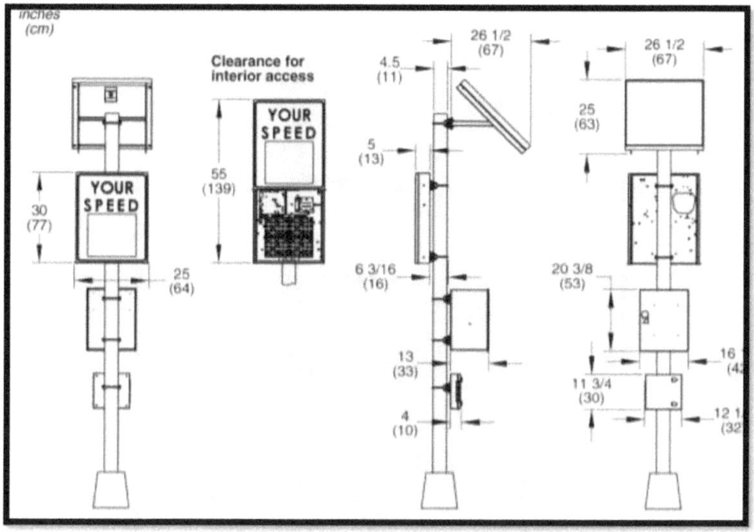

Figure 2-14: System dimension of the Pole-mount RSS developed by Wanco Inc [19].

SafePace 100 developed by Trafficlogix

The 3rd system to be investigated was the SafePace 100 developed by Trafficlogix [20]. This system consists of three sub-systems: the seven-segment LED display unit, A k-band CW radar module, and a 20 W solar power system. The display unit has 208 LEDs with a maximum brightness of 87964 lux. The radar module has an accuracy of $\pm$ 1 km/h and can obtain returns from up to 200 m. The maximum speed the radar can display is 99 km/h. The power system can provide 12V DC at 720 mA [20]. The dimensions in L x W x H are 584.2 x 88.9 x 736 in mm, and the system display sign was shown Figure 2-15.

Figure 2-15: SafePace 100 developed by Trafficlogix [20].

The SafePace 100's system specifications are summarized in Table 2-10.

Table 2-10:The SafePace 100's system specifications [20].

Parameters	Specifications	Units
Supply voltage	12	V
Carrier frequency (F_c)	24.125	GHz
Max Doppler shift	4.4	kHz
DR	N/A	dB
SQNR	N/A	dB
Operating current	720	mA
Power output (EIRP)	20	dBm
Antenna beam-width (Az x El)	38 x 45	°
Sidelobe levels (Az x El)	N/A	dB
Temperature rating	-40 to 85	°C
Physical dimensions (L x W x H)	584.2 x 88.9 x 736	mm

MSPM 2 developed by Monitor systems

The next system to be studied was the MSPM 2, developed by Monitor systems [21]. The system consists of three sub-systems. The first sub-system is a k-band CW radar sensor that operates at 24.150 GHz, $\pm$ 100 MHz. The system has an accuracy of $\pm$ 2 km/h, for targets travelling at speeds between 5 km/h and 99 km/h. The second sub-system is an LED seven-segment display with an unspecified maximum brightness. The last sub-system is a 50 W solar panel power system that can provide 12 V DC at 2 A. The system is enclosed in a stainless-steel enclosure. The metric model's dimensions in terms of height x width x depth are 749.3 x 584.2 x 88.9 in mm. The system was shown in Figure 2-16.

Figure 2-16: MSPM 2 developed by Monitor systems [21].

TC-400 portable radar speed sign developed by Radarsign

The radar system studied in this example was the TC-400 portable radar speed sign developed by Radarsign [22]. The TC-400 consists of four sub-systems. The radar module, the LED display module, and the battery system.

These sub-systems are housed in an aluminium enclosure that had a silver powder coat finish. The chamber's physical dimensions in terms of L x W x H are 577.85 x 60.33 x 412.75 in mm. The radar module is a k-band CW system with an operating frequency of 24.125 GHz, $\pm$ 50 MHz.

The system's speed measurement accuracy was within $\pm$ 1.6 km/h for speeds between 8 km/h to 205 km/h. The system is powered by two 12V DC Ni-MH battery packs that can provide 4.5 Ah each. The system has a two-digit LED display visible up to 137.16 m away in a day and night conditions [22]. TC-400's system specifications were summarized in Table 2-11.

Table 2-11: TC-400's system specifications [22].

Parameters	Specifications	Units
Supply voltage	12	V
Carrier frequency (F_c)	24.125	GHz
Max Doppler shift	4.4	kHz
DR	N/A	dB
SQNR	N/A	dB
Operating current	720	mA
Power output (EIRP)	20	dBm
Antenna beam-width (Az x El)	12 x 12	°
Sidelobe levels (Az x El)	N/A	dB
Temperature rating	-40 to 71.1	°C
Physical dimensions (L x W x H)	577.85 x 60.33 x 412.75	mm

The system was shown in Figure 2-17.

Figure 2-17: TC-400 portable radar speed sign developed by Radarsign [22].

The system weighs 11.78 kg without the battery packs and can operate between -40 °C and 71.1 °C. The radar can store data of up to 56 million vehicle detections and retain it for 12 months.

Speed Monitor F developed by MPH industries

The last example to be studied was the Speed Monitor F developed by MPH industries. This system consists of three sub-systems: an approach only CW radar sensor and a 460 mm two-digit super bright LED speed display, which was visible at up to 360 m in both day and night conditions. The last sub-system was a 12 V DC power system. All these sub-systems are housed in a 14-gauge steel enclosure that has waterproofing [23].

Table 2-12: Speed Monitor F's system specifications [23].

Parameters	Specifications	Units
Supply voltage	12	V
Carrier frequency (F_c)	24.125	GHz
Max Doppler shift	4.4	kHz
DR	N/A	dB
SQNR	N/A	dB
Operating current	720	mA
Power output (EIRP)	13.01	dBm
Antenna beam-width (Az x El)	12 x 12	°
Sidelobe levels (Az x El)	N/A	dB
Temperature rating	-40 to 71.1	°C
Physical dimensions (L x W x H)	910 x 230 x 760	mm

The radar sensor is a k-band CW system with a 360m detection range and a velocity accuracy of ± 2 km/h. It transmits a signal with a maximum power of 25mW, and the antenna beam-width at -3dB is 12° in both azimuth and elevation [23]. The system weighs 48kg, and its dimensions in terms of L x W x H are 910 x 760 x 230 in mm. The system was shown in Figure 2-18.

Figure 2-18: Speed Monitor F developed by MPH industries [23].

Summary

Table 2-13 is a summary of the commercial systems reviewed in this section.

Table 2-13: Summary of the technical specifications of the most common radar speed signs imported by the South African companies mentioned in this study.

Parameter	Huston Radar	Wanco	Trafficlogix	Monitor systems	Radarsign	MPH industries
				Manufacturers		
Product name	PNL10	Pole-mount RSS	SafePace 100	MSPM 2	TC-400	SM F
Mode	CW	CW	CW	CW	CW	CW
Velocity Estimation Accuracy	$\pm$ 1.6 km/h	$\pm$ 3.2 km/h	$\pm$ 1 km/h	$\pm$ 2 km/h	$\pm$ 1.6 km/h	$\pm$ 2 km/h
Detection Range	90 m	305 m	200 m	N/A	366 m	360 m
Max speed	159 km/h	161 km/h	99 km/h	199 km/h	205 km/h	99 km/h
DC power	18 V	12.8 V	12 V	12 V	12 V	12 V
Solar	No	65 W	20 W	50 W	No	20 W
Two lane system[5]	No	Yes	No	N/A	yes	No
Display Brightness	11000 cdm	21000 cdm	12,000 cdm	N/A	700 cdm	1000 cdm
Sensor Frequency[6]	24.125 GHz $\pm$ 50 MHz	24.125 GHz	K-band	24.150 GHZ $\pm$ 100 MHz	24.125 GHz, $\pm$ 50 MHz	K-band
Weight	450g	35kg	33 kg	48 k g	35.78 kg	48 kg
Reliability[7]	2 years	2 years	2 years	2 years	2 years	2 years
Cost[8]	R12945.6	R70560	R52416	R65404.8	R72952	R78984

The manufactures of these radar-based traffic calming systems do not disclose the detection rates of their systems.

The experimental and commercial systems reviewed thus far have their merits and shortcomings. It was observed that the experimental designs were relatively cheaper than the commercial structures. However, these systems did not include dedicated power solutions, which means there are no solar-powered options or battery packs that provide all-day power. These systems do not include dedicated enclosures to enable commercial deployment. The inconsistency in the data provided about these systems does not lead to reliable conclusions about these systems'

---

[5] Detects two vehicles simultaneously
[6] IEEE definition of K-band is a frequency band from 18 to 27 GHz
[7] Warranty for product
[8] All cost excludes shipping and the prices shown calculated using the exchange rate of R15.21/$

detection performance. However, a 90% detection rate was observed for some of the solutions studied.

The significant advantage that the commercial systems have over experimental designs is that they are readily deployable in most urban conditions. They have solar options for autonomous power requirements as well as all-day battery capabilities. However, these systems are prohibitively expensive. Thus, the proposed solution must have a combination of advantages from both the experimental and commercial systems; this includes a detection rate equal or greater than 90% and can have specifications the enable deployment and autonomous power capabilities. All these strengths, while having a cost below R20k.

2.3   Summary

This chapter introduced experimental and commercial radar systems. The technical specifications detailed in the investigation helped to enable a proper understanding of the current technological landscape. This investigation was done to contextualize the user requirements stated in Section 1.3. The requirements include the typical velocity estimation accuracy, the detection range, and costs associated with such a system's manufacturing. These system requirements were used in making the design decisions of the proposed system. Observations made from Table 2-7and Table 2-13 led to the conclusion that the most dominant architecture used for speed calming is CW at K-band. The k-band has an advantage of being an unlicensed spectrum, this means it does not require a license to operate in that frequency band [24].

The next chapter details the theoretical considerations for radar systems modelling, showing the advantages and disadvantages of using specific radar architectures and how these systems are modelled mathematically.

Chapter 3

Theoretical Considerations for Radar Modelling

This chapter provides the theoretical considerations for understanding and designing a radar system for our application. The chapter first outlines the basic principles that govern radar. Then an in-depth description of the different classes of radar waveforms and their typical uses follows. With the above in mind, the basic building blocks of a typical CW radar system were described to formulate meaningful technical requirements for our radar system.

3.1 Radar Principles

In order to design an appropriate radar speed sign and obtain the requirements, the basic principles of radar must be understood. The basic working principle entails a radar antenna transmitting an electromagnetic signal called the transmitted signal. The transmitted signal illuminates the target that is at a specific range from the radar. The transmitted signal then reflects from the target and an attenuated version of the transmitted signal, which was known as an echo signal, travels back to the receiving antenna to the radar processor as illustrated in Figure 3-1. The radar processor may extract the properties of the target from the received echo signal, such as the direction of travel, the speed of the target and range of the target depending on the specific waveform and radar configuration used.

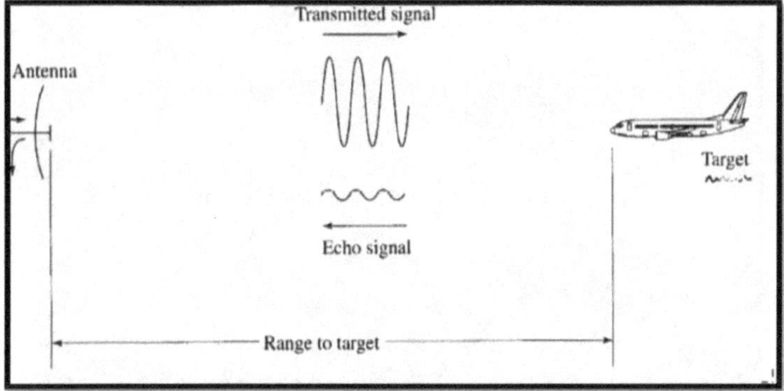

Figure 3-1: Radar principle. The radar transmits a signal through the antenna, which was reflected by a target back to the antenna [25].

3.2   Radar Waveforms

In Radar engineering, there are mainly two types of waveform classes [26]. These are continuous waveforms (CW) and pulsed radar systems, but there exist multiple sub-classes of these systems as detailed in Figure 3-2 [9]. In the case of pulsed radars, this includes fixed frequency pulse, pulse to pulse modulation, frequency agility and intrapulse modulation.

In the subclass of intra-pulse modulation for pulsed systems, there are other modulation techniques such as phase modulation as well as frequency modulation, which are further sub-divided into bi-phase modulation, e.g. Baker-codes, polyphase modulation, e.g. Welti-code and Frank-code. In the case of frequency modulation, there was linear FM, e.g. sawtooth as well as non-linear FM, e.g. triangle, pseudo-random, noise modulated, stepped frequency as well as sinusoidal. There was also frequency shift keying under the frequency modulation sub-class for both CW and pulsed radar systems. All these systems work on a principle of manipulating phase and frequency properties of signals to either find the range of a target or the velocity of a target. These techniques are also used for sidelobe reduction and to reduce the blind range of the radar [27].

CW radar includes multiple frequency CW and an unmodulated CW radar system. In an unmodulated CW radar system, the transmitter continuously transmits a monotone signal whilst simultaneously receiving the echo. Since the radar continuously transmits and receives signals, CW radars often employ two antennae, one for transmission and another for receiving; this allows the radar to alleviate transmitter coupling into the receiver. A CW radar sensor system was illustrated in Figure 3-5.

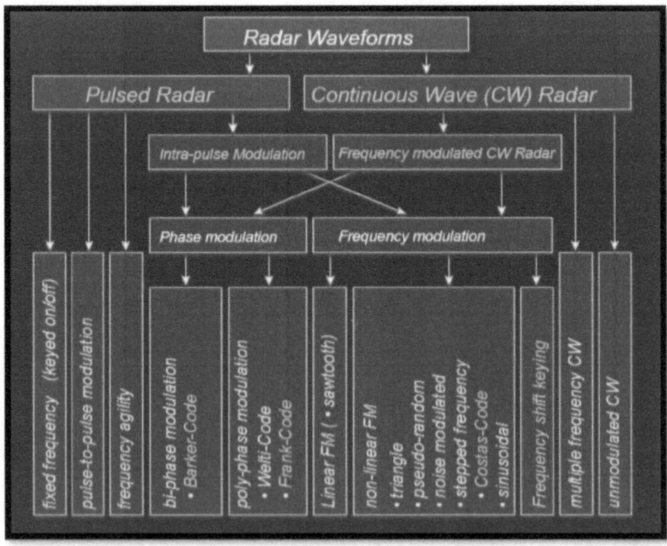

Figure 3-2: Radar waveforms [9]

In order to measure the velocity of a target at a given time as well its range, the CW system can be augmented to make use of a frequency-modulated continuous waveform (FMCW).

The introduction of frequency modulation allows for greater signal bandwidth, which increases range resolution capability. The range of the target may be obtained by calculating the Fourier transform (FT) of the beat frequency, f_r. The beat frequency f_r is the frequency difference of the received signal and the transmitted signal and is shown in Figure 3-3 [27].

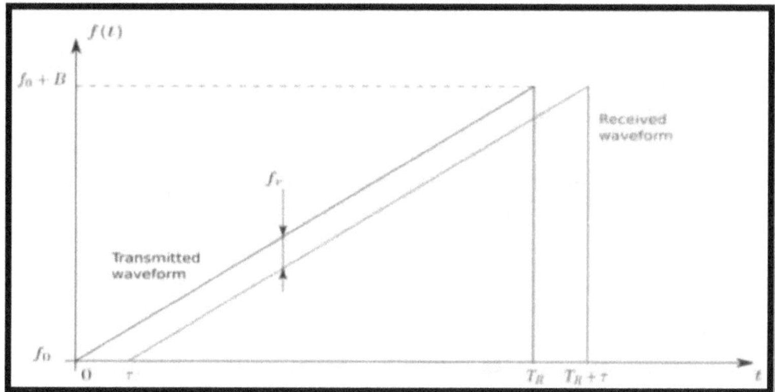

Figure 3-3: Frequency modulated waveform [27].

Figure 3-4 illustrates how a CW signal can be modified to achieve a pulsed radar system by periodically turning on and off the power amplifier that exists between the splitter and antenna. The resultant signal is a series of pulses illustrated on the bottom half of Figure 3-4.

Pulsed radar systems typically employ high powered transmitters, depending on the application; thus, insulation is necessary to protect sensitive receiver electronics. They are used to obtain the ranges of targets, but pulse-Doppler radars are used to obtain the velocity measurements of targets as well, they require coherent transmitters, and the receiver must have a high instantaneous dynamic range [26]. Other applications of both CW and pulsed radar systems include imaging systems [26].

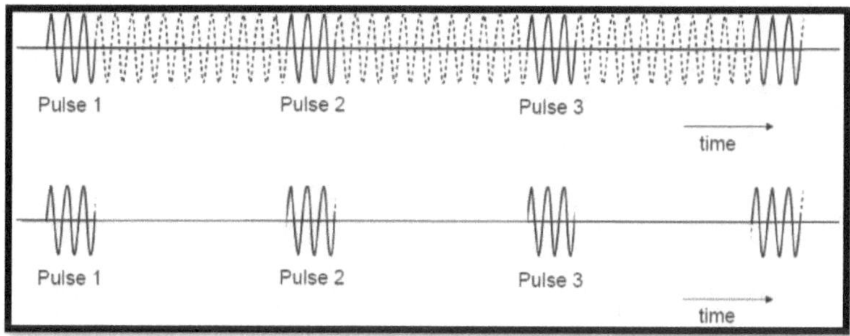

Figure 3-4: Pulsed signal generation. A pulse being generated from a continuous waveform [26].

3.3   CW Radar System Overview

The first element OCS1 in the functional block diagram, shown in Figure 3-5 represents an oscillator [28]. Which is used to generate a signal of frequency f, and that signal was amplified by element AMP1 to a desirable level for transmission. The gain of the amplifier depends on the required transmit power level. The next element in the chain was a signal splitter, denoted by SPLTR1. This element allows the amplified signal to pass through whilst also providing an attenuated version of the original signal. The high amplitude signal was transmitted via the element ANT1, which represents an antenna.

The second antenna ANT2, receives echoes of the transmitted signal after it was reflected from the target. Echoes are amplified by a low noise amplifier LNA1. This amplifier has a low internal noise floor which makes it well suited as it does not add significant additional noise, and this results in a higher signal to noise ratio. Signal to noise ratio was discussed in detail in Section 3.5.

The low amplitude signal from the splitter was mixed with the amplified echo from LNA1. The element that was responsible for mixing the two signals is called a mixer, denoted by MXR1. The resultant signal was filtered and amplified by a video amplifier [28].

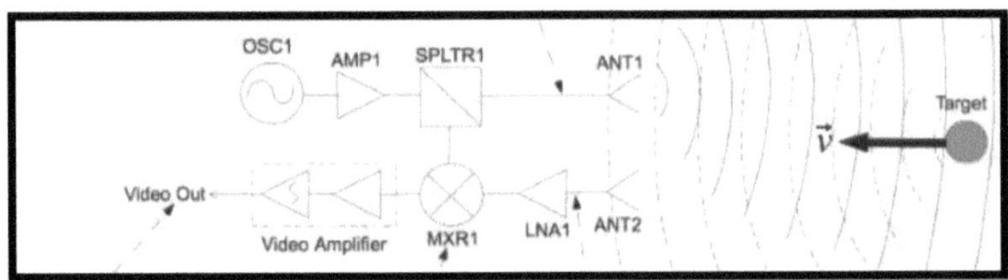

Figure 3-5: CW radar principle. A block diagram of a CW radar sensor was shown with how
CW propagate and interact with a target travelling towards the radar [28].

CW systems are usually relatively low power with a typical transmission power of a few watts or less and are thus used predominantly for short-range applications. CW radars can be used to find the radial velocity of a target through Doppler processing.

3.4   CW Radar Signal Modelling

In Section 3.2, CW waveforms were briefly introduced. In this section the properties of CW waveforms are explored in detail.

A single unmodulated sinusoid, transmitted at carrier frequency f_0 can be modelled as:

$$x(t) = \text{A} \cos(2\pi f_0 t) , -\infty \leq t \leq \infty \qquad\qquad 3\text{-}1$$

Where:

A: Signal amplitude in linear units

t: time in seconds (s)

As can be seen from Equation 3-1, the signal model was continuous for all time. The frequency difference of the signals expected at the mixer output, MXR1, also represents the target Doppler frequency which can be estimated by Equation 3.2. When an object of velocity v moves towards the radar as depicted in Figure 3-5 [29], the Doppler frequency was described as:

$$f_{Dopp} = 2f_0 \times \frac{v}{c} \times cos(\alpha)$$

3-2

Where:

f_{Dopp}: Doppler or differential frequency in Hertz (Hz)

v: Velocity of moving object in meters per second (m/s)

c: Speed of light in meters per second (m/s)

α: Angle of the direction of the object motion relative to the radar in radians (rad)

3.5 Radar Cross Section and Signal to Noise Ratio

Radar Cross Section (RCS) is a target parameter that quantifies the EM scattering phenomenology in radar technology. It has units in m^2 typically expressed in dB scale as dBsm.

RCS is a function of target viewing angles relative to the radar transmitter and receiver antennae, as well as the radar frequency, polarization of the incident EM wave and the target size, shape and material properties. In summary, *"RCS is a measure of not only the how much of the incident EM wave is reflected from the target but also how much of the incident wave intercepted by the target and how much is directed back toward the radar's receiver"*[26].

In this study, the targets of interest are average vehicles with sizes similar to the VW Golf V, VW Beatle, an Audi A4 estate and a Fiat Ducato van studied by Schipper et al. [30], who found that these vehicles have RCS values between -12dBsm and 25dBsm when measured with HH and VV polarization at 23-27 GHz this is consistent with Swerling 1 [26].

The average RCS at a front viewing angle or 0° was found to be a mean value of 12.5 dBsm [30]. 12.5 dBsm was the value used in this study; this means that targets of interest in this study are attributed with an RCS of 12.5 dBsm and are considered to be non-fluctuating targets for simple parameter calculations [26].

According to Richards et al. [26], a target that is non-fluctuating relates to an object with a constant RCS when viewed at every viewing angle, at every range and at all times. This approximation is only limited to modelling the proposed system when designing the system at the maximum detectable range, a more practical model must be considered such as Swerling 1, which assumes a scan to scan decorrelation from randomly distributed scatterers, of which none were dominant [26]. Once assembled, the practical performance detection performance must be characterized to see how accurately the model approximated the system.

Signal to Noise Ratio (SNR) is the ratio between the received signal power and the noise power at the receiver. In order to be able to detect targets from the distances similar to those of systems outlined in Table 2-13, a sufficient target SNR must be measurable by the radar system.

The theoretical SNR may be calculated using the following [26]:

$$SNR = \frac{N\sigma P_t G_t G_r \lambda^2}{(4\pi)^3 R^4 P_n}$$

3-3

Where:

N: Number of pulses

σ : mean RCS in m^2

P_t: Transmitted power in Watts

G_t: Transmit antenna gain in linear units

G_r: Receive antenna gain in linear units

λ: Wavelength in meters

R: Target range in meters

P_n: Noise power in Watts

The noise power at the receiver can also be calculated as follows:

$$P_n = kT_0BF \tag{3-4}$$

Where:

k: Boltzmann's constant (1.3807 x 10 ^-23 Joules per kelvin)

T_0: Standard temperature in Kelvin

B: Instantaneous receiver bandwidth in Hertz

F: Noise figure of the receiver subsystem (linear units)

In order to obtain the anticipated noise power, the noise figure required can be obtained using the following [31]:

$$F = S_r - SNR_{min} - 10\log_{10} kT_0B \tag{3-5}$$

Where:

SNR_{min}: Minimum Detectible SNR in dB

S_r: Receiver sensitivity in dBW/Hz

Since the radar module is an integrated system as described in Section 1.7.1, the amplifiers cannot be directly probed to obtain the noise figure. Thus, experimental methods must be used to determine the observed SNR using the system [26].

$$SNR = \frac{A_s^2}{\sigma_n^2} \tag{3-6}$$

Where:

A_s^2: Signal amplitude is proportional to the power of the signal

σ_n^2: Noise variance is the equal to the noise power assuming Gaussian white noise

The observed SNR must be defined as the ratio of the total signal power to the noise power.

$$SNR_{obs} = \frac{A_s^2}{\sigma_s^2} \tag{3-7}$$

Λ includes both the signal and noise contributions, Equation 3.6 does not account for the noise component found in the observed amplitude; thus, the observed SNR is considered as the signal-plus-noise to noise ratio. The actual SNR of a target can be calculated as the expected value of the observed SNR [26], that is

$$SNR = E\{SNR_{obs}\} - 1 \qquad\qquad 3\text{-}8$$

Where E {} denotes the expected value also known as the mean

3.6   Analogue to Digital Conversion

Most modern systems make use of digital signal processing techniques in order to process target returns. One of the advantages of working with digitized signals is improved SNR because signals that have been digitized with sufficient resolution can be amplified to achieve desired amplitude and the noise can be digitally filtered out [26]. In order to exploit these properties, the analogue to digital conversion principle must be understood. This section explores this topic in greater detail.

Digital signals are obtained by discretizing analogue signals. This process involves sampling and quantizing the signal, as illustrated in Figure 3-6. This entails sampling the ADC's input voltage and then holding this voltage for the duration of the conversion. The *Sample and Hold* (S/H) circuit performed this task and is located directly at the input of the ADC. In typical ADCs, the S/H briefly opens its aperture window and captures the input voltage on the rising edge of the clock signal. It then closes the aperture window to hold its output at the newly acquired level. This output level is updated at every rising edge of the clock input of the ADC [32].

Depending on the chosen quantization scheme, a numerical value is then assigned to the voltage level present at the output of the S/H. This is referred to as quantization. The quantizer searches for the nearest corresponding value to the amplitude obtained by the S/H. This is chosen from a fixed number of possible values that cover its complete amplitude range. Since the quantizer cannot search from an infinite number of possibilities, it was restricted to a limited set of potential values. The size that this set is linked to spans the dynamic range of the quantizer and the set was equal to two raised to the nth power, where N was referred to as the number of bits.

The number of bits used by the ADC to encode the digital values is also referred to as the resolution of the ADC.

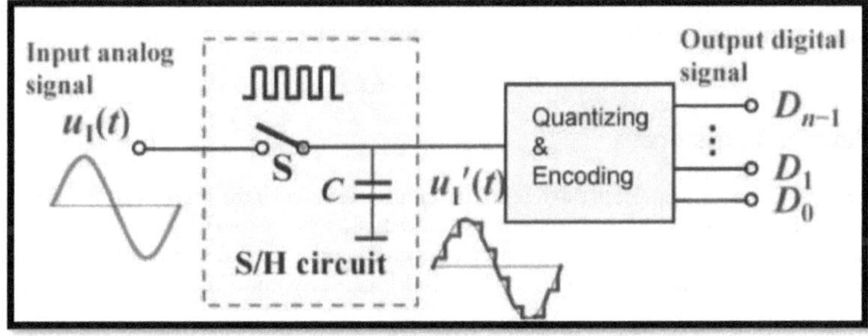

Figure 3-6: Analogue to digital conversion process [32].

Using the modelled CW waveform signal, the process illustrated in Figure 3-6, can be modelled as

$$x_s[t] = x(t)p(t) \qquad\qquad 3\text{-}9$$

Where the sampling function $p(t)$ can be modelled as a periodic impulse train with the mathematical representation of

$$p(t) = \sum_{n=-\infty}^{\infty} \delta(t - kT) \qquad\qquad 3\text{-}10$$

k was an integer that establishes the time position of each time signal.

The ADC must have sufficient resolution in order to accurately represent the In-phase and Quadrature components of the signals produced by the radar sensor; this was of importance since harmonic distortions will arise in the signal spectra as a result of quantization errors. Quantization errors are a result of the ADC not having sufficient resolution to reconstruct the input analogue signals digitally [9]. It can be seen from Table 3-1 that the higher resolution, the lower quantization errors.

Table 3-1: ADC characteristics [32].

Resolution (in bits)	Range	± Range	Quantization Error	Maximum Sampling Rate
24	16,777,216	± 8,388,608	± 0.000003%	200 KHz
16	65,536	± 32,768	± 0.0008%	250 MHz
14	16,384	± 8,192	± 0.003%	400 MHz
12	4,096	± 2,048	± 0.012%	1.8 GHz
10	1,024	± 512	± 0.05%	2.2 GHz
8	256	± 128	± 0.2%	3 GHz
6	64	± 32	± 0.8%	6 GHz

An advantageous property of analogue to digital conversion was the improved signal to quantization noise ratio (SQNR). This was a result of the relationship between the SQNR and resolution of the ADC which was illustrated by [33] for a completely sinusoidal signal.

$$(SQNR)_{dB} = 1.76 + 6.02n \qquad\qquad 3\text{-}11$$

Where n was the resolution of the ADC. The assumptions made to reach Equation 3.11 was that the signal was sinusoidal as was stated in Equation 3.1 and that the quantizer was selected to cover a voltage range of $\pm\frac{1}{2}$ without saturation, the bias term 1.76 changes as the waveshape of the test signal changes as shown in example 8-3 by Ziemer et al. [33].

The ratio of the largest representable magnitude to the smallest nonzero magnitude was called the Dynamic Range (DR) by Richards et al. [26], which can be represented as

$$DR = 6.02n - 6.02 \;\; dB \qquad\qquad 3\text{-}12$$

3.6.1 Fourier Transform

Once the signal has been sampled, the frequency characteristics of the signal needs to be analysed in order to obtain the Doppler frequencies. This task can be performed using the Fourier transform. This algorithm converts the signal from the time domain into a frequency domain representation. A comparison of the number of multiplications required using the FFT algorithm was shown in Table 3-2.

Table 3-2: Comparison of the number of real multiplications for the DFT and FFT [33].

Number of Points	DFT	FFT	Speed Factor
2	8	4	2
4	48	16	3
8	224	48	5
16	960	128	8
32	3968	320	12
64	16128	768	21
128	65024	1792	36
256	261120	4096	64
512	1046528	9216	114
1024	4190208	20480	205
2048	16769024	45056	372
4096	67092480	98304	683

The FFT is an algorithm that enables the radar to obtain the spectral approximation of the signal. The longer the FFT (more time samples), the finer the frequency resolution of the result, as expressed by Equation 3-13.

$$\delta_f = \frac{f_s}{K} \qquad \qquad 3\text{-}13$$

This result leads to a trade-off with time resolution, as it requires a longer time signal to obtain a finer resolution frequency spectrum.

3.7 Detection of Targets

The most fundamental task of radar is target detection, which involves processing the radar data and deciding on whether the information acquired represents interference only or if targets are present in the data [26]. Threshold detection is the process of deciding by means of a threshold signal level, whether a signal was noise or a reflection from a target. This concept was illustrated in Figure 3-7.

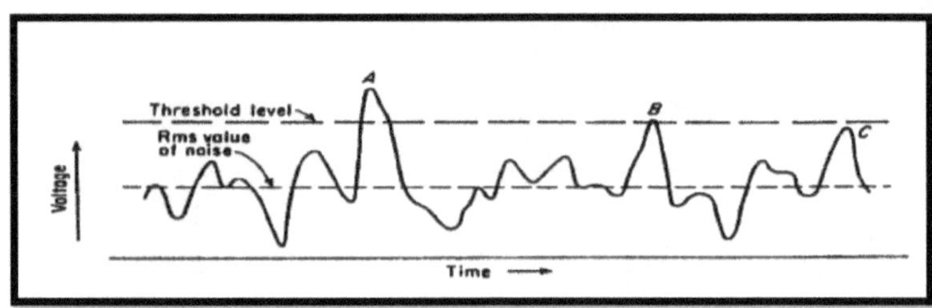

Figure 3-7: Concept of threshold detection. The illustration depicts a signal in volts, a threshold level is used to determine at what level can a signal be considered a target echo [27].

Figure 3-7 illustrates that the process of detecting a target on the basis of the signal voltage was a statistical process characterised by a probability of detection P_D, that is usually less than unity and a probability of false alarm P_{FA}, that is greater than zero [26].

In a radar measurement tested for the presence of a target, one of the following hypotheses can be assumed to be true:

- H_0, the measurement was the result of interference only.
- H_1, the measurement was the combined result of interference and echoes from a target.

The first hypothesis was known as the null hypothesis, denoted by H_0, and the second as H_1.

Specialised detection strategies are not commonly revealed in industry from fear of electronic counter measures being deployed to prevent detections [25]. This was the case with devices studied in Table 2-13, and therefore the most common detection algorithm was employed in this study. According to Richards et al. [26], radar detection algorithms are typically designed using the *Neyman-Pearson criterion*; this is an optimization strategy that fixes the P_{FA} that will be allowed by the detection processor and then maximizes the P_D for a given SNR.

Considering that the target[9] of interest was a vehicle which was assumed, for this study, as a non-fluctuating target in Gaussian noise, the probability of false alarm can be modelled as follows [26]:

$$P_{FA} = \exp\left(-\frac{T^2}{\sigma_n^2}\right) \qquad\qquad 3\text{-}14$$

The optimum threshold was thus:

$$T = \sigma_n\sqrt{-lnP_{FA}} \qquad\qquad 3\text{-}15$$

The above result provides the rule for setting a threshold at the output of a linear detector.

The theory suggests that the probability of detection of a non-fluctuating signal in Gaussian Noise for a linear detector in terms of Marcum's Q function was [26]:

$$P_D = Q_M\left(\sqrt{\frac{2\hat{m}^2}{\sigma_n^2}}, \sqrt{\frac{2T^2}{\sigma_n^2}}\right) \qquad\qquad 3\text{-}16$$

Where:

---

[9] RCS ~ 12.5 dBsm

$\tilde{m}$: Target component of the signal echo.

This detection strategy was used for simple modelling for system parameters and to obtain sensible technical requirements. In later chapters, a more suitable detection strategy must be investigated to produce sensible and suitable results.

3.8   Radar Measurements

Once the target has been detected, the next goal would be to measure the velocity of the target. In Section 3.2, it was established that Doppler radars could determine the velocity of the desired target. Section 3.4 showed the modelling of a typical CW system, and Equation 3.2 described how the velocity of an object could be obtained from using such a system.

The quality of the measurement of a quantity such as the velocity was characterized by its precision and accuracy. Accuracy is the difference between the measured value and the actual value whilst "precision characterizes the repeatability of multiple measurements of the same quality, even when the accuracy is poor" [26], as illustrated in Figure 3-8.

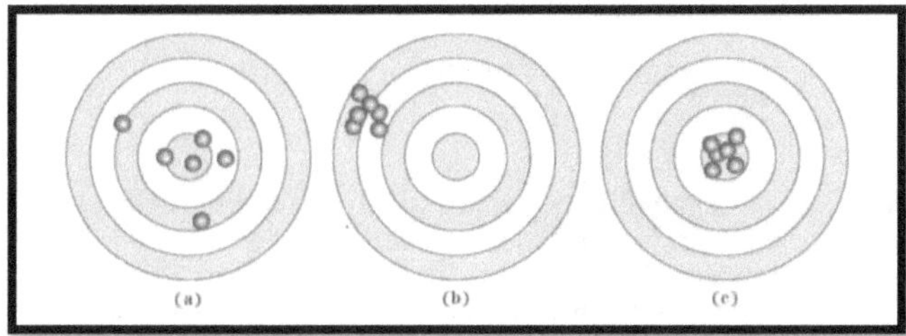

Figure 3-8: Illustration of accuracy and precision in target shooting. (a) Accurate but imprecise (low error mean but high standard deviation). (b) Precise but inaccurate (low standard deviation but high mean error). (c) Precise and accurate (low standard deviation and low mean error) [26].

In order to obtain statistically sound results, a large enough sample of measurements must be obtained using the same measurement procedures. Measurements can be evaluated in terms of precision and accuracy by obtaining the mean and standard deviation of the measurement errors for each set of measurements that were obtained from the same experiment.

3.8.1   Parameter Estimation

The radar measurement process has only one objective, to estimate the characteristics of an object of interest from its reflected signal echo. The parameter of interest in this study was the Doppler shift of the target. Hence, it is beneficial to discuss the general idea of an estimator and what precision is achievable.

Consider an observed signal $y(t)$ that is a sum of a target component $s(t)$ and the noise component $w(t)$, assuming that the noise is Gaussian:

$$y(t) = s(t) + w(t) \qquad\qquad 3\text{-}17$$

The signal $y(t)$ may be a function of one or more parameters α_i. In this study, the parameter is the Doppler shift. Thus, the objective is to estimate this parameter given a set of observations of $y(t)$ by use of an *estimator*. Suppose $y(t)$ is sampled to produce a vector of N observations.

$$Y = \{y_1 y_2, \ldots\ldots\ldots, y_N\} \qquad\qquad 3\text{-}18$$

Noise energy in the signal *y(t)* means that the vector Y is random and depends on the parameter α. Therefore a conditional probability density function (PDF) *p(Y/ α)*. Thus, the estimator would be defined as *f* for the parameter α based on the data Y.

$$\hat{\alpha} = f(Y) \qquad\qquad 3\text{-}19$$

Since Y is random, the estimate $\hat{\alpha}$ is also a random variable with a probability density function with a mean and variance.

The properties most desired in an estimator are that it is unbiased and that it is consistent. This means that the expected value of the estimate equals the actual value of the parameter. The variance of the estimate decreases to zero, as more measurements become available [26]. This means:

$$E\{\hat{\alpha}\} = \alpha_i \text{ (Unbiased)} \qquad\qquad 3\text{-}20$$

$$\lim_{N \to \infty} \{\sigma_{\hat{\alpha}}^2\} \to 0 \text{ (Consistent)} \qquad\qquad 3\text{-}21$$

There are many types of estimators, but since this study considers the noise to be zero-mean Gaussian noise with variance σ_n^2. The estimator used is maximum likelihood (ML) estimator. It is standard practice to use an ML estimator as its form is often relatively easy to determine, and since the noise is considered Gaussian, it is an optimum estimator [34]. In order to obtain the minimum achievable variance, which is the (square precision) the Cramér-Rao lower bound (CRLB). The CRLB for measuring the Doppler shift for the signal described by Equation 3.1 with M measurements, assuming the initial phase and amplitude are unknown is shown by [34] as:

$$\sigma_f \geq \frac{\sqrt{3}}{\pi} \frac{\delta_f}{\sqrt{SNR_f}} \qquad\qquad 3\text{-}22$$

Where :

δ_f: Frequency resolution

SNR_f: Frequency domain Signal to noise ratio

Equation 3-22 states that the precision of the frequency estimate is proportional to the frequency resolution divided by the square root of the applicable SNR. This relationship means the precision of the Doppler measurements for a given Doppler resolution improves when the SNR of the measurement large; this also means that for a very fine Doppler resolution, the target is required to have a high SNR to be measured precisely, i.e. repeatable high accuracy velocity measurements require high SNR.

3.9 Summary

In this chapter, the theoretical considerations for radar modelling to were put forward, allowing for the CW architecture to be chosen and modelled. The radar equation was also was explored in this chapter, paving the way for accurate technical requirement specifications to be developed.

Considerations such as the RCS non-fluctuating targets were made to make calculations of specific design decisions to be more accessible. The assumption that a small vehicle's mean RCS when viewed at $0°$, is 12.5 dBsm was formed by Schipper et al. [30].

The next chapter details the formulation and analysis of these system requirements to formulate the application test procedures (ATPs). The ATPs would help determine the success or the failure of the system when measured against a predetermined set of metrics.

Chapter 4

System Requirement Development and Methodology

4.1 User Requirement Motivation

The user requirements detailed in Section 1.3 outline the basic tenants of a radar-based traffic calming solution. The systems that were investigated in Chapter 2 had either CW, FMCW or pulsed a systems architecture, but the CW architecture was found to be the most prevalent system architecture used and in Chapter 3 the mathematical considerations associated with this architecture were explored. In this section, the motivations behind the user requirements provided in Section 1.3 are given.

The first requirement was specific to radar detection capabilities. This investigation was aiming to obtain the maximum distance small vehicles can be detected. Small vehicles may be classified as small hatchbacks and sedans, such as the Ford Fiesta or the Toyota Yaris.

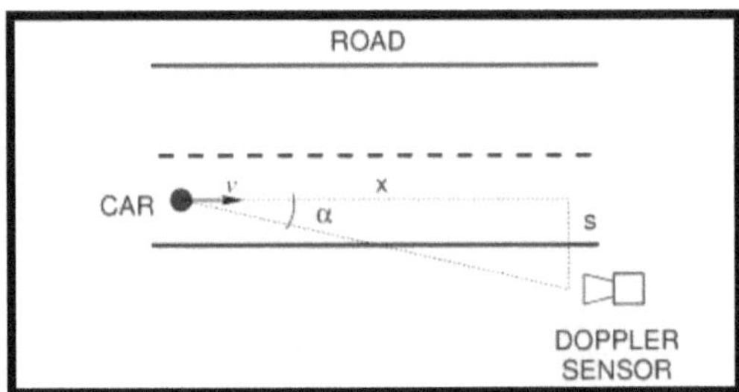

Figure 4-1:Typical radar location relative to the road and car [15].

Since the typical radar location would be beside the road, the radar must be placed at the distance S beside the road such that the angle formed by the line connecting the car and the radar sensor, α, is reduced: this is to ensure that the discrepancy between the actual velocity v, and measured velocity, vp is eliminated. Alimenti et al. [15] concluded that vp and v could be assumed to be equal when the distance between the car and the radar sensor is greater than 3S, which in real terms is 15 m. Therefore, alpha must be between 0°-20° for a maximum tolerable discrepancy of 5% or ±2 km/h for cars travelling at the maximum speed allowed at the CSIR campus of 40 km/h;

for these reasons, it can be assumed that the radar illuminates the front profile of the vehicle at all times while the car is approaching the radar.

To obtain a suitable radar detection distance, the radar must satisfy a scenario where a vehicle travelling at high speeds must be considered. According to Hoole et al. [3], a safe reaction time of a driver travelling behind a vehicle is 2 seconds [3]. With this in mind, it can be concluded that for a car travelling at 60 km/h, which is the typical speed limit for urban conditions [3], the driver would travel 33.33 m before they appropriately react to prompt from the radar. Given a 10% buffer distance of 3.3 m added as a safety factor [3], the maximum detection range becomes 36.63 m or 40 m after rounding up. The CSIR campus roads are short and narrow, and since this system would only be used for campus traffic calming, 40 m was considered sufficient.

A correlation emerges when comparing the cost of systems with longer detection ranges; the larger the detection range, the higher its cost. Thus, the maximum detectable range of 40 m for small vehicles was investigated to keep costs low. Large vehicles have a large RCS, which means they can be detected at more considerable distances.

The second requirement pertains to the accuracy of the velocity measurement of the proposed system. It was required to be "comparable or better than current commercial and experimental systems" for similar applications. In Table 2-7, the average speed accuracy is $\pm$ 1.245 km/h, and the average speed for commercial systems, in Table 2-13, is $\pm$1.9 km/h. Therefore, a "comparable" speed measurement accuracy for a system such as the proposed system is $\pm$2 km/h.

The third requirement pertains to the visibility of the display and the intelligibility of the displayed velocity estimate. A safe reaction time of a driver travelling behind a vehicle is 2 seconds according to Hoole et al. [3] in the official k53 manual; since 40 m is the maximum distance, a detection must occur, the driver of the vehicle must be notified of their speed when the radar detects that the vehicle is breaching the speed limit at that range.

It can be reasonably extrapolated that a driver requires at least 2 seconds to react to a speed sign instruction to slow down. The maximum speed allowed at the CSIR campus was 40 km/h; assuming a car travels at 60 km/h or 16.667 m/s, the driver will have 2.4 seconds to reduce their vehicle speed before they pass the speed sign.

The fourth requirement pertains to the reliability of the system. The standard warranty of electrical components of this system must be at least 2 years; this was a typical warranty offered by most of the commercial systems in Table 2-13, and in ensuring that the proposed system was comparable to the reliability of these systems, it must be designed with components with the same warranty.

The fifth requirement pertains to the system's continuous operation; since the speed sign is based on radar technology, it can operate in the daytime and night-time. The energy considerations of the system are also brought to light. The system must have a constant and reliable supply of power. Therefore, the system might have to rely on battery technology and solar technology to ensure that it works day and night.

The last user requirement pertains to the total cost of the system, including labour. The proposed system must deliver the same/comparable performance to commercial systems while remaining affordable. Since affordability is relative, the budget given for the components that make up the system was R15k, while the labour costs being R5k. On account of the system described in this report being only a data-collection system, the emphasis was placed on obtaining quotations for non-essential components, and only essential components were purchased. Components such as the radar sensor, an essential component in the system, cannot be sufficiently represented by a system other than itself. Meaning, no other system can produce results that are characteristic of it. Specifications of the computational systems may be used to determine whether the proposed system is capable of the real-time nature of this project. Thus, a data-collection system was only used to determine the quality of the data produced by the radar sensor.

4.2   Technical and Non-technical Requirement Formulation

In his section, the system requirements were derived using information contained in Section 4.1. The first user requirement quantifies the detection range. It was concluded that 40 m was a feasible detection range when accounting for the reaction times of the motorist, as stated by Hoole et al. [3]. Assuming the radar is placed beside the road as depicted in Figure 4-1 and the maximum speed being 60km/h, and since the minimum velocity at the CSIR campus was 20 km/h, the requirement becomes:

4.2.1   Requirement 1

- The system shall be able to detect targets up to a distance of 40 m from the radar when travelling at a velocity between 20 km/h to 60 km/h

The technical requirement specified above gives rise to the question of the quality of the detections made by the radar. The quality of these detections is determined by the number of false detections present in a random sample of measurements.

The maximum velocity that this system must be able to measure was 60 km/h, it was found that the Doppler frequency induced by an object travelling at 60 km/h measured by a radar system with a transmit frequency of 24 GHz is 2666.72 Hz.

A sampling frequency of 10 kHz would satisfy the Nyquist criterion [26]. Thus, the performance of the system depends on the number of false detections observed for every 10 thousand samples collected.

In Section 3.7, it can be seen that the detection performance of a radar detector was characterized by the probability of false alarm and the probability of detection. The probability of detection was influenced by both the probability of false alarm as well as the signal to noise ratio.

 Unfortunately, not all of the commercial and experimental systems that are featured in Table 2-7 and Table 2-13 disclose the detection performance of their systems, those that do place the detection rate at 90%, which that means from a measurement set with ten true detection observations, nine of those measurements were correctly identified as true detection measurements that contain the target.

In order to increase the probability of the system of correctly identifying true detection measurements and dismissing false positives, two parameters must be chosen to this end. The first being the probability of detection P_D and the second is P_{FA}. Which means given a set of random measurements, the likelihood of a true detection measurement being positively identified as a detection in that set is about P_D. The probability of false alarm P_{FA} is the likelihood that from the same random set, a non-detection measurement may be falsely characterized as an accurate detection.

The receiver operating curve (ROC) simulations are required to find suitable radar performance objectives. Assuming 10k samples are obtained every second, the ROC shows which combination of P_{FA} and P_D corresponds to which SNR based on the Neyman-Pearson linear detector.

Table 4-1: Number of false alarms expressed as P_{FA}

Error in 10k measurements	Probability of false alarm P_{FA}
0.0000000000000001	1e-20
0.001	1e-7
0.01	1e-6
0.1	1e-5
1	1e-4
10	1e-3

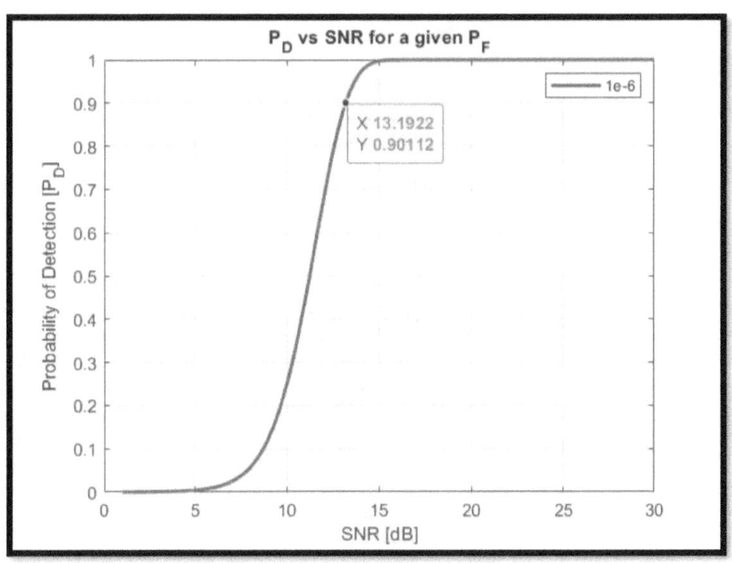

Figure 4-2 : P_D vs SNR for a given P_F [27]

It can be observed from Figure 4-2 that the minimum SNR that would result in a P_D greater than 90% is anything above 13.19 dB. The complimentary P_{FA} is 1e-6; this means that after one million measurements, there is a statistical probability that one measurement may be a false detection.

4.2.2 Requirement 2

- The system shall have a probability of false alarm of 1e-6 and the probability of detection of 90%, for targets at 40 m travelling at a velocity between 20 km/h to 60 km/h.

The second user requirement is with regards to the velocity estimation of the proposed system. A comparable velocity estimate accuracy was found to be within $\pm$ 2 km/h. This velocity accuracy was quantified at 40 m. This technical requirement can only be fulfilled if the velocity resolution of the system is finer than 2 km/h, this was dictated by the sampling frequency as well as the number of FFT points. Therefore, the requirement becomes:

4.2.3 Requirement 3

- The system shall have a velocity measurement accuracy of $\pm$ 2 km/h at 40 m, when the target travels at a velocity between 20 km/h to 40 km/h.

The third user requirement pertains to the visibility of the radar speed sign at 40 m. It has been established that the driver of an incoming vehicle will have sufficient response time when travelling at a maximum observable speed of 60 km/h. Thus, a display device with sufficient luminous efficiency in the day and night-time must be chosen for this application. Pu et al. [35] demonstrate in the study on the legibility of LED traffic guide signs in urban tunnels that 20 cm thin stroke LEDs with a luminance of characters of 150 cd has an average visual cognition of 70 m [35]. Thus, the requirement becomes:

4.2.4 Requirement 4

- The system shall have a speed display with a luminous intensity greater than 150 cd for visibility and legibility in the day and night-time at a minimum distance of 40 m.

The fourth user requirement pertains to the reliability of the system since it would require 2 years to prove that the system would indeed last that long. It was concluded that for the system to have the reliability comparable to that of commercial systems, the warranty of all the components used must be at least two years or more.

The fifth user requirement involves the system's power considerations; the systems reviewed in Table 2-7 and Table 2-13 have different power considerations. Whilst the experimental systems are mostly powered by a 5 V DC source or less; commercial systems have very different power needs. Most commercial systems are powered by a combination of 12 V DC supply and solar panels. The solar panels supply power during the day and charges the battery after use of the battery from the previous night. The average weight of these systems was 48 kg when considering systems that come with two deep cycle battery packs. The average weight of a 100 A.h battery is 28 kg [36].

Thus, the requirements become:

4.2.5 Requirement 5

- The system shall have uninterrupted power during the day and night through a combination of solar and battery power.

4.2.6 Requirement 6

- This system shall have a combined weight of less than 48 kg; this includes the solar panel, the enclosure with its contents, i.e. DSP, battery, radar sensor and display.

The last user requirement dictates that the total cost of the system and the assembly must cost R20k or less. The R15k must be reserved for the component costs and the assembly of the system by a skilled technician must equate to R5k. The component costs include the cost of the enclosure with all the protective equipment that such a system would require. Thus, the requirement becomes:

4.2.7 Requirement 7

- The system shall have a total cost to manufacture of R20k, with the components and the enclosure costing less than R15k and R5k for labour.

4.3 Preliminary Design Considerations and Tests

This section outlines the development of the preliminary tests. The outcomes of these preliminary tests will provide information that is important in the fulfilment of the requirement specifications. Precisely, the first three requirements.

Requirement 1 relates to the system's ability to make detections at a maximum distance of 40 m, while the vehicle is in motion. The velocities that are of interest are between 20 km/h and 60 km/h. The detection capabilities of a radar system are directly influenced by the average power received at the radar as dictated by the radar range equation shown in Equation 4-1.

$$P_r = \frac{P_t G_r G_t \lambda^2 \sigma_e}{(4\pi)^3 R^4} \qquad\qquad 4\text{-}1$$

Where

P_t is the peak transmitted power in watts.

G_t is the gain of the transmit antenna in linear units.

G_r is the gain of the receive antenna in linear units.

σ_e is the mean RCS of the target in square meters.

R is the range from the radar to the target in meters.

The SNR of the radar system, discussed in Section 3.5, provides a measure of the target energy present in the signal relative to the noise; this means that the sensor used to obtain this data must have a sufficiently high transmit power and high receiver sensitivity. The sensor bandpass filter must also have sufficient bandwidth in order to capture the frequencies that correspond to the velocities of interest.

The selection involves making a detailed comparison between radar transceiver modules typically used for this application, using device specifications from data sheets and internet sources.

After the selection of the appropriate radar sensor, the first step would be to determine the correct operation of the radar module as specified by the manufacturer. The next step would be to obtain spectrograms of the radar location scene, to get an estimate of the noise and interference energy of that scene.

Once the correct operation is confirmed, and the typical noise intensity of the scene determined, the next step would be to investigate the experimental SNR; by measuring the signal energy reflected from the vehicle as it approaches the radar. These measurements are required when determining the change in SNR when the vehicle is at different distances.

4.4 System Requirement Analysis and Application Test Procedure Development

This section outlines the development of the application test procedures (ATPs) based on requirements stated in Section 4.2. The theoretical development of these requirements was guided by the technology survey summarized in Table 2-13.

Fulfilling Requirement 1 requires a positive detection to be declared while the vehicle travels towards the radar at speeds between 20 km/h and 60 km/h, starting at 40 m from the radar.

An experiment to fulfil Requirement 2 would be to have the vehicle travel towards the radar sensor from 40 m, the vehicle must either travel at either 20 km/h or 60 km/h. Then a spectrogram from the recording must be analysed in order to observe if the performance of the system fulfils this requirement.

Requirement 3 pertains to the accuracy of the radar-based traffic calming system. Radar speed signs are usually placed just in front of physical speed calming measures such as humps; to prevent cars from travelling at dangerous speeds in an area that humans and animals use cross the road.

Consider a motorist travelling at a speed of 45 km/h at a section of road with a speed limit of 40 km (which is the limit at the most business campuses including the CSIR) and the radar had an error of $\pm$ 5 km/h, if the radar detects a speed of 40 km/h, the radar will not display a warning for the motorist to slow down.

This would result in the vehicle to drive over the hump at high speed, causing an uncomfortable jerking motion and would accelerate the wear and tear of the suspension of the vehicle. The worst-case scenario is that a non-motorised user getting injured by the vehicle while crossing the road.

In order to fulfil this requirement, observations detailing the velocities of vehicles travelling at constant speeds must be obtained using the assembled system. Then the velocity estimations must be closely examined to determine if the velocity of the car indeed does match that estimated by the radar, and to what degree does the estimate differ from the ground truth. If the deviation is within $\pm$ 2 km/h, then Requirement 3 would be fulfilled.

Requirement 4 requires obtaining a LED display that has a significant luminous intensity. This device can be found through a detailed comparison of LED displays that have a luminous intensity greater than 150 cd. This quantity is also stated in the datasheets of these devices. Thus, the appropriate selection of this module would also ensure that this requirement is fulfilled.

Requirement 5 relates to the reliability of the power supply as well as redundancies that will ensure constant power. Therefore, there must be an accounting of the total power needs of the system and must include the energy requirements of the radar module, ADC, DSP and the speed display. Then an energy solution that includes a combination of batteries and solar power must be presented.

Requirement 6 alludes to the total weight of the system. This requirement can be satisfied by adding all the weights of the components making up the system including the enclosure. The design of a structure that can carry such a load must be presented to ensure the successful fulfilment of this requirement.

Requirement 7 is a critical aspect of this project as the financial incentive of creating an alternative radar speed sign is the central theme of this study. This is measured by the total cost of the bill of materials. Different quotations must be sourced for each component in order to obtain cost-effective alternatives to those presented in Table 2-13; this also includes quotations from electricians with skills to assemble such a device.

The following is a table summarising the preliminary tests and experiments used to obtain data to fulfil the requirements.

Table 4-2: Summary of the preliminary tests and experiments.

Requirement	Reason for experiment	Experimental set up	Experimental procedure
1,2	To determine if the radar was fully operational.	The radar must be placed under laboratory conditions.	A hand must be swag 60cm from the radar to obtain a sinusoidal pattern in the spectrogram as advised by the manufacturer [37].
1,2	To obtain the experimental Noise intensity of the environment that the radar will be placed in.	The radar shall be placed outside and shall face the direction where it will be permanently placed.	The radar data will be collected for a full 10 seconds. Then using this data, the mean noise intensity will be calculated using samples that are from the noise and from the clutter. Using these values, the noise energy may be deduced and the clutter + noise intensities can also be deduced.
1,2	To obtain the experimental SNR.	The radar must be placed 40m from the measurement starting position, such that the broad sight of the beam was directly illuminating the target and there was no angle between the target and the beam.	A small car must travel a distance of 40m towards the radar. The car must maintain a constant speed throughout the measurement to enable the calculation of SNR as it approaches the radar.
1	To detect car from a distance of up to of 40m from the radar, when travelling at a speed between 20km/h to 60km/h	The radar must be placed at 40m from a hump, such that the broad sight of the beam is directly illuminating the target and there was	The vehicle must travel 39m towards the radar and stop 1m away from the radar.

			no angle between the target and the beam.	
3	To obtain the accuracy of the system.	The radar must be placed 40m from the starting position of the car, such that the broad sight of the beam is directly illuminating the car front and there was no angle between the car and the beam	An electric car must travel at a constant speed of 20, 40 km/h. Since experiment would be done on campus grounds campus security do not permit an experiment whereby a motorist would travel at a constant speed of 60 km/h. An electric car was able to electronically maintain a constant speed through the cruise control feature.
2	To measure the required 60km/h and then correct car speed to 20km/h.	The radar must be placed 40 m from the starting position of the car, such that the broad sight of the beam is directly illuminating the car front.	The car must approach the radar the radar at 60 km/h from 40 m way and then slow down to 20 km/h.

Requirements 4 through 7 do not have experiments to prove their validity; the reason for this is that the proposed system would not be built. Only the specifications of the individual components must meet the requirements for them to be fulfilled. In other words, data sheets must be analysed to obtain the quantities that match the requirement specifications, and this will be deemed proof of concept.

4.5   Methodology

The radar system design must be able to fulfil the requirements stated in Section 4.2. The following is a description of how to design and assemble the radar.

4.5.1   Identify Suitable Radar Hardware

The first step in the design of this radar system is to create a block diagram of the radar hardware chain. This step allows for an overview of all the hardware considerations that should be made. The second step in the design process involves stating the requirements each sub-system or component aims to fulfil or partially fulfil. This step leads to identifying possible radar transceiver modules; these transceiver modules must have specifications that are suitable for the speed calming application. Then after choosing a suitable module and stating the reasons for the selection, the next step is to identify an appropriate ADC to meet the specifications that would lead to the fulfilment of the requirements stated in Section 4.2.

The succeeding steps are the identification of items such as various DSPs, display units, as well as power units. These components of the system must all be chosen appropriately to enable the fulfilment of the relevant technical requirements stated in Section 4.2. However, before a power unit may be identified and chosen, a power budget that details the energy needs of a completed radar system must first be created.

4.5.2 Radar Signal Processing Algorithms

In order to code the various radar signal processing algorithms, first proper software engineering principles must be followed. The first step requires a thorough description of the processes contained in the algorithm. Then using this description of the signal processing chain, proper pseudocode, or a flow chart should be derived. Once the pseudocode/flow chart has been outlined, an appropriate programming language must be chosen in order to execute the program efficiently and appropriately.

4.5.3 Sub-system Integration and Testing

The first step in the integration of the various sub-system is the testing of each sub-system to ensure correct operation. Then the integration of the radar sensor hardware, signal processing algorithms of the radar prototype to form the representative radar-based traffic calming system.

Doppler vs time spectrograms must be created by using data acquired from the sensor in controlled laboratory conditions, as well as real-world conditions to ensure that the system is working as intended; this includes measuring vehicles moving at different velocities and obtaining the signal to noise ratio of the system at different ranges. There should also be a process to test the software of the system.

Thus, simulated data must be used to test the correct operation of the chosen detection strategy. All issues associated with incorrect operation or integration of the hardware must be detailed and addressed.

4.5.4 Hardware Specification for the Proposed System

The last part of this process is to obtain quotations and data sheets with specifications of the power components, enclosures and DSPs, to be used in the final design of the proposed system. The system to be designed will form a blueprint of what an affordable traffic calming must entail.

4.5.5 Project Constraints

This aim of this project is not to develop a commercially ready Doppler radar; instead, it is to design a system that has specifications comparable to commercial systems in both accuracy and quality. The system itself must have components not costing more than commercially available radars but have the same function. The specified system in this project will be prototyped only using a laptop and a data collection device similar to that of commercial systems.

4.6 Summary

This chapter provides the development of the system requirements. It also details the development of application test procedures (ATPs). These procedures enable benchmarking the performance of the developed prototype against the radar speed signs in Table 2-7 and Table 2-13. The next chapter details the design of the proposed system. Simulations are also used to help quantify the correct specifications of system components required to fulfil the requirements stated in Section 4.2.

Chapter 5

System Design and Simulations

5.1   System Design

Table 2-7 and Table 2-13 highlighted the radar speed sign and detection characteristics. These systems fulfil the requirements stated in Section 4.2 of this study. The following section is a detailed design that led to the prototyping of a device that was comparable in technical performance to the experimental and commercial systems investigated in Section 2.2. Unfortunately, most of the systems investigated in Table 2-7 and Table 2-13 did not disclose their detection capabilities, but the systems that had disclosed this information produced detection rates between 90%- 95% [13] [8]. The proposed system must have a probability of detection at 90% and a probability of false alarm of 1e-6. This system must also fulfil all other requirements stated in Chapter 4.

The common building blocks of a radar speed sign typically consists of the following:

- Radar sensor
- Analogue to Digital Converter (ADC)
- Digital Signal Processing (DSP) block
- Display module
- Power source

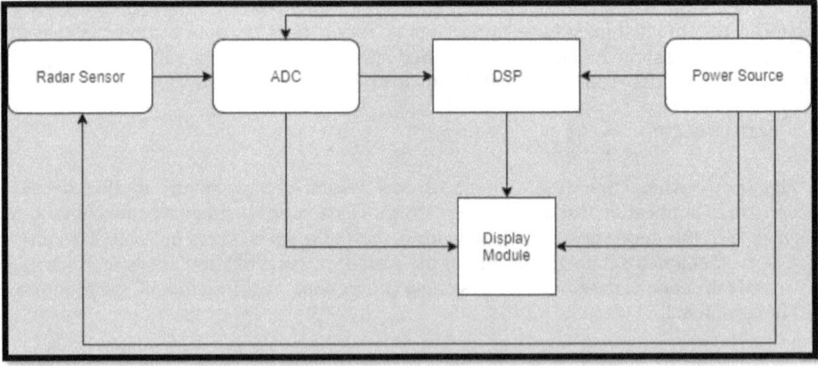

Figure 5-1: Radar Block diagram. Simplified system block diagram of a radar speed sign which demarcates the project scope.

Radar sensor/Module: : The radar sensors used in the devices found in Table 2-13 are integrated systems which house both transmit and receive antennae, continuous waveform (CW) generators modulators and demodulators among other components. The output of the radar module is In-phase and Quadrature phase sinusoids with typical voltage levels in the order microvolts. Thus, a need for a suitable radar sensor is a requirement.

ADC: Signals received from the radar module are required to be converted from analogue to digital form in order to be suitable for radar signal processing.

DSP Module: The DSP then processes the digitized data to make the necessary detections and speed measurements. This information is then converted into a format that can be readily displayed.

Display module: This sub-system takes the data produced by the DSP module and presents it in a human-readable form. The driver of the vehicle must be able to see the presented information at a sufficient distance for them to be able to correct their driving behaviour when the need arises. The display module enables the engineer to verify the quality of data obtained from the ADC during testing and validation.

Power module: The power modules must be able to provide for the energy needs of all sub-systems. Most have different varying power needs. Thus, the power source needs to be adaptive.

5.1.1 System Overview

The radar system consists of two main components, the radar hardware and radar signal processing. The radar hardware consists of the following sub-systems. The radar transceiver module, the ADC, the digital signal processor, the power system and the display unit. This was illustrated in Figure 5-1

The hardware configuration illustrated in Figure 5-1 shows a generic configuration allowing the proposed device to be powered by one source, in reality the power source must be specified to suit the power requirements of every component, thus the power source shall be specified last.

The radar sensor is selected first because the properties of the data collected by this sensor dictate the parameters of the desired ADC. These parameters include sufficient resolution, as stated in Section 3.6 , the resolution of an ADC allows the signal to be accurately reproduced and determines the dynamic range of the system.

The system that provides the signal processing must be specified; this system should be real-time, so the DSP of this system should be able to process the data near real-time.

The radar signal processing consists of the following sub-processes. The first sub-process captures the radar data. The next sub-system reconstructs the complex signal and extracts its frequency components. The detection algorithm then follows, and the measured speed reading is displayed. This process is illustrated in Figure 5-2.

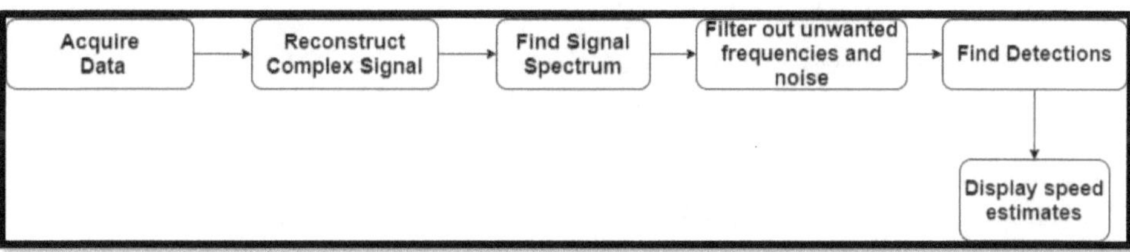

Figure 5-2: Radar signal processing overview.

The system illustrated in Figure 5-2 is a generic flow diagram of a radar signal processing algorithm. The first sub-system requires reading of the data to the systems memory, then the data must be manipulated in order to reconstruct the complex signal since sampling of I and Q channels are done separately in this system. The complex signal has phase information that contains the Doppler frequencies.

An FFT must be done to obtain the Doppler frequencies. A detection algorithm will then be applied to distinguish between noise and possible target detections. Once a detection decision has been made and the association confirmed, the target Doppler frequency will then be converted into a velocity estimate, and then be displayed.

The display should be visible from a distance far enough that the driver will have sufficient time to correct their behaviour.

The first step in assembling the radar system involves obtaining a suitable radar transceiver module that will be able to contribute to satisfying Requirements 1, 2, 3 and 7.

5.2   Radar Sensors

The considerations that are made in this section aided in the selection of a radar sensor. It is not a requirement to build a radar from first principles since numerous solutions already exist. However, Requirements 1, 2, 3 and 7 were carefully analysed, and specifications for an appropriate radar module were derived.

The following guidelines were derived from Requirements 1, 2, 3 and 7 as they relate to specifications of the radar sensor:

- An operating range of 40 m
- A relatively high output power
- A relatively low input power
- A minimum and maximum observable speed of 20 km/h and 60 km/h.
- A relatively low-cost

The following parameters are known variables based on investigations in Section 4.1:

- Detection Criterion
    - $P_D = 0.9$
    - $P_{FA} = 1e\text{-}6$
    - $SNR_{min} = 13.19$ dB
- Operating parameters
    - Target RCS ~ 12.5 dBm
    - $R_{max} = 40$ m
    - Operational Frequency $= 24$ GHz
- Resolution
    - Doppler resolution < 2 km/h

In order to distinguish which radar module would aid in satisfying the above-stated guidelines, the experimental and commercial radar systems from Table 2-7 and Table 2-13 in Section 2.2 the literature survey was carefully examined. What was apparent in almost all these systems was the use of a k-band radar module. The reason is that the k-band of the electromagnetic spectrum was allocated for general use by ICASA, meaning that it does not interfere with critical instruments used in the military and commercial systems [24]. The advantages of k-band also include relatively small compact antennas as well as relatively low atmospheric losses of 0.1 dB/km and high bandwidth [26]. Thus, it was concluded that the radar modules that would be compared would be k-band systems. There are many radar manufactures that produce k-band radar systems. This study will focus on modules manufactured by Innosent GmbH since the majority of commercial systems reviewed make use of their systems. This German company also complies with ICASA regulations [24]. They also have local suppliers; therefore, timely procurement should be expected.

5.2.1 CW Radar module architecture

In this section, the radar module architecture is examined, and the different internal components and their functions are discussed, in Section 3.3 the general architecture of CW radars is discussed in detail and in this section the nuances' of the 24 GHz sensors that are used for traffic calming are unpacked.

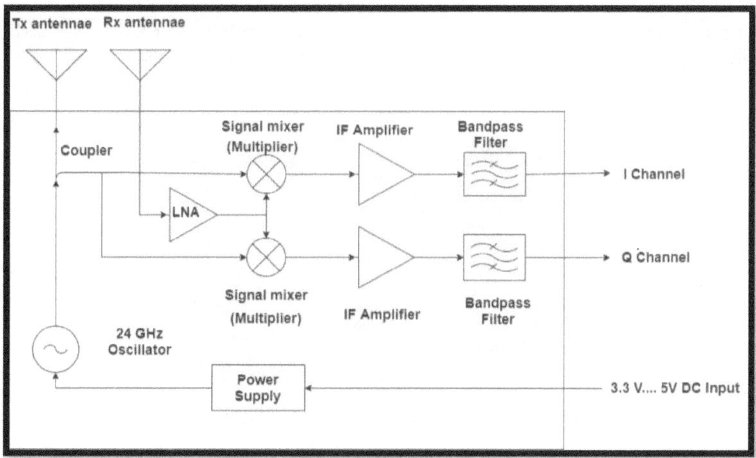

Figure 5-3: 24GHz CW radar module schematic [37]

Figure 5-3 shows the internal architecture for a typical 24 GHz CW radar module. The transmission antennae are fed with a 16 – 20 dBm CW signal at 24 GHz; the 24 GHz oscillator produces this signal. A 0.2 -0.5 W power supply powers the oscillator and the module is powered by a 3.3 – 5 V DC input. The coupler taps of a low amplitude copy of the Tx signal which will be used for demodulation. The received signal is routed through the low noise amplifier and mixed with the low amplitude Tx signal in order to obtain an intermediate frequency (IF) signal. The IF signals finally go through a bandpass filter to eliminate unwanted frequencies, the output of the bandpass filters are the I/Q components of the baseband signal [37]. The baseband signal is the output of this module and all the signal processing outlined in Figure 5-2 is performed on the digitized version of this signal.

5.2.2 Radar Module Desirable Specifications

This section outlines the specific parameters that the prospective radar module must have in order to be considered suitable for the proposed system. The prospective radar module must have enough bandwidth to be able to capture frequencies between 888 Hz and 2.67 kHz, which are the Doppler frequencies corresponding to objects moving at 20 km/h and 60 km/h respectively. The antenna beam-width must be large enough to illuminate a single lane; the side-lobes must also be low enough to avoid unwanted detections from targets, not within the main lobe. The device must have a relatively small footprint and an extensive detection range. The device must also have a cost lower than R5k as this was the maximum budget as per user specification.

5.2.3   Radar Module Comparison and Selection

The first module to be investigated is the IPS-355 which shown in Figure 5-4. This module is a small and low power, CW radar sensor, which consumes 0.345 Watts of power and has an output power (EIRP) of 12.7 dBm. This sensor has an operating range of up to 40 m and has a full beam-width of 70° in azimuth and 36° in elevation at -3 dB beam-width. This module can simultaneously detect six targets [38]. This module has side-lobe levels of 13 dB both in the horizontal plane in the vertical plane [39].

The physical dimensions of this module are 8.3 x 44 x 30 mm in height x length x width. This module has a bandwidth of DC- 1 MHz. Operating temperatures are between -40°C and 60°C.

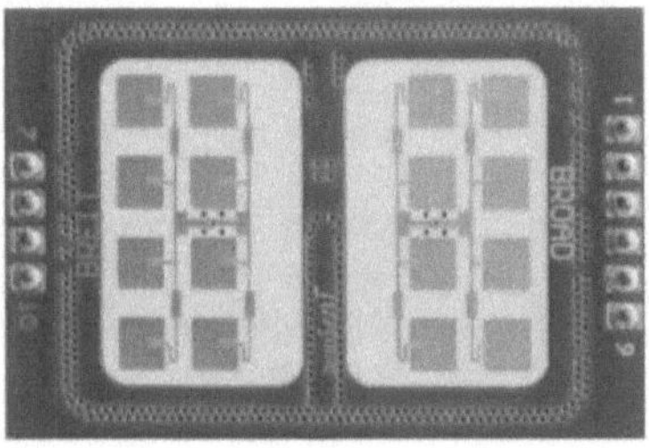

Figure 5-4: IPS-355 K-Band FMCW Radar sensor [39].

The next module to be investigated is the IPS-354 radar module, which was shown in Figure 5-5. This module is a small and low power, CW radar sensor, which consumes 0.345 Watts of power and has an output power (EIRP) of 12.6 dBm. This sensor has an operating range of up to 30 m and has a full beam-width of 45° in azimuth and 38° in elevation at -3 dB beam-width. This module has side-lobe levels of 15 dB in the horizontal plane and 20 dB in the vertical plane [39]. The physical dimensions of this module are 8.3 x 44 x 30 mm in height x length x width. This module has a bandwidth of DC- 1 MHz. Operating temperatures are between -40°C and 60°C.

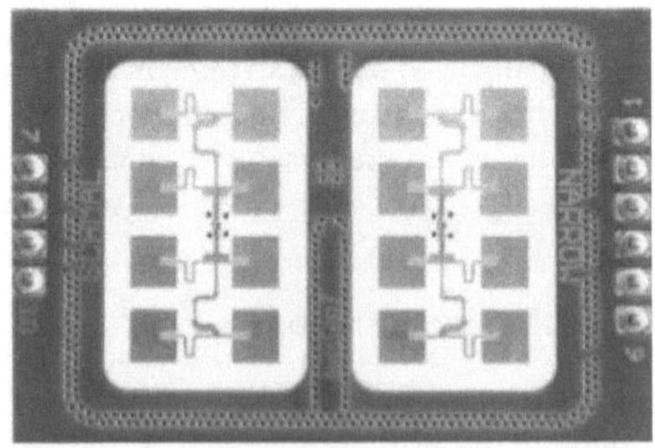

Figure 5-5: IPS-154 CW K-Band Radar sensor [40].

The IPS-937 radar sensor shown in Figure 5-6 is a CW transceiver that is mainly used for traffic monitoring applications. This module has two 4 x 4 patch antenna arrays. The module consumes 0.34 W of power; the transmit power is 20 dBm. This module has a range of operation of up to 350 m. The low noise amplifier (LNA) has 40 dB of gain. The bandwidth is between 30 Hz and 10 kHz.

This module has a beam-width at -3 dB of 33° and 33 ° in azimuth and elevation respectively. The side-lobe levels of this module are -25 dB in both azimuth and elevation. The outline dimensions in mm are 9 x 61.6 x 37 in height x length x and width. This module can operate at temperatures between -40°C and 85°C [41].

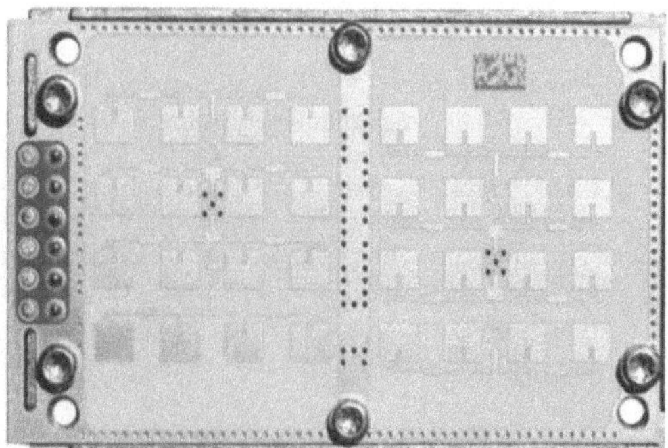

Figure 5-6: IPS-937 K-Band Radar sensor [41].

The IPS-280 radar sensor shown in Figure 5-7 is a CW Doppler transceiver; this module has a transmit power of 20 dBm and a bandwidth of DC to 1 MHz. This module can only obtain returns from one vehicle at a time since it has a beam-width at -3 dB of 9° and 18° in azimuth and elevation respectively. The power consumed by the module was 0.221 W; the module has a transmit power of 20dbm and has a range of operation of a 100 m. The module has the following dimensions in mm of 10.2 x 70 x 65.8 in height x length x width. This device can operate at temperatures between -40°C and 85°C [41].

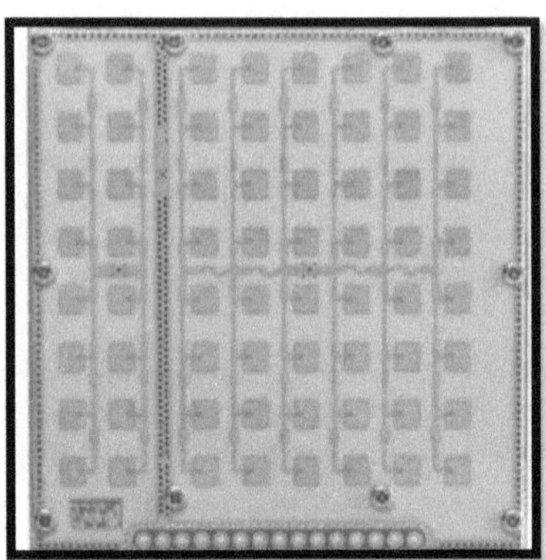

Figure 5-7: IPS-280 K-Band CW Doppler Radar sensor [42].

The IPS-144 radar sensor shown in Figure 5-8 is a long-range traffic monitoring transceiver, it has a range of operation of up to 150 m. This module consumes 0.42 W of power and the transmit power was 20 dBm. The sensor beam-width can only fit a maximum of two targets at a time and at -3 dB has a beam-width of 12° and 25° in azimuth and elevation respectively. The side-lobe levels are 20dB in both azimuth and elevation. The bandwidth was between 50 Hz and 20 kHz. This module has 53 dB of antenna gain. The outline dimensions in mm are 11 x 65.8 x 65.8 in height x length x width. This module can operate at temperatures between -20°C and 60°C [43] [42].

Figure 5-8: IPS-144 K-Band CW Doppler Radar sensor [43].

The last sensor investigated is the IPS-154 k-band CW Doppler sensor shown in Figure 5-9. This module is typically used as a door opener and for industrial applications. This module has a beam-width at -3dB of 45° and 38° in azimuth and elevation respectively. This radar sensor has on output power of 20 dBm and gain of 20 dB. The module has a bandwidth between DC and 50 kHz. The power consumed by this module is 0.263 W. The outline dimensions in mm are 8.3 x 44 x 30 in height x length x width. This module can operate at temperatures between -30°C and 60°C [37] [43]. This module has an operating range of 40 m.

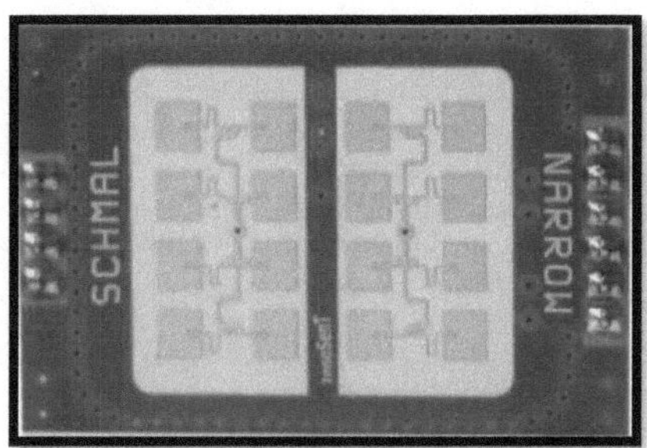

Figure 5-9: IPS-154 K-Band CW Radar sensor [37]

Table 5-1 gives a detailed comparison between the various specifications of the radar sensors.

Table 5-1: Summary of specifications and costs of fixed pole speed sign sensors.

		Fixed pole Radar speed signs sensors					
Parameter	units	IPS-355	IPS-354	IPS-937	IPS-280	IPS-144	IPS-154
Power input	W	0.35	0.35	0.34	0.22	0.42	0.25
Power output	dBm	12.7	12.6	20	20	20	20
Full beam-width @-3dB	° azimuth	70	45	33	9	12	45
	° elevation	36	38	33	18	25	38
LNA gain	dB	20	20	40	52	53	20
Bandwidth	Hz	DC – 1 M	DC – 1 M	30 -10 k	DC-1M	50 - 20 k	DC-50k
Architecture		CW	CW	CW	CW	CW	CW
Side-lob levels	dB azimuth	-13	-15	-25	-40	-20	-15
	dB elevation	-13	-20	-20	20	-20	-20
Temperature	°C	-20 and 60	-20 and 60	-40 and 85	-40 and 85	-20 and 60	-30 and 60
Range	m	40	30	350	100	150	40
Maximum detectible vehicles		6	2	1	1	2	6
Outline dimensions	mm length	8.3	8.3	61.6	10.2	11	8.3

	mm height	44	44	9	70	65.8	44
	mm width	30	30	37	65.8	65.8	30
Cost[10]	R	1450	1350	1444	5200	8000	R850

In Table 5-1 the different radar specifications were summarised, the most desirable specifications were highlighted in red. All the radar sensors surveyed have at least one desirable trait, but the sensor that had the most desirable specifications was the IPS-154.

Thus, the IPS-154 was the transceiver of choice. The factors that contributed to this choice include, the relatively low cost of the module. The sensor also can obtain target returns from up to 40 m. This sensor also can observe speeds from 1 km/h to 1125 km/h, and the bandwidth is not too high to be affected by high-frequency interference. This module has side-lobe levels of -15 dB in azimuth and -20 dB in elevation, which is comparable to side-lobe levels with more expensive sensors. The cost of this system was substantially lower than all the proposed sensors; this was the main contributing factor in the decision to choose it.

The minimum SNR for a $P_D = 0.9$ at a $P_{FA} = 1 \times 10^{-6}$ is 13.19 dB using the specifications of the IPS-154 the following theoretical SNR may be obtained.

Table 5-2: SNR calculation using IPS-154 specifications

Parameter	Symbol	Value	units
Radar Output Power (EIRP)	P_t	20	dBm
Minimum SNR	SNR_{min}	13.19	dB
Receiver Bandwidth	B	50	kHz
Transmit frequencies	f	24	GHz
Wavelength	$\frac{c}{f} = \lambda$	0.0125	m
Mean RCS	σ	12.5	dBsm
Gain (Tx and Rx)	$\Rrightarrow$	9.5	dBi
Range	R	40	m
Standard temperature	T_0	295	K
Boltzmann's constant	k	1.3807×10^{-23}	J/K
Receiver Sensitivity	S_r	-121	dBm/Hz

Using Equation 3-3, $SNR = \frac{\sigma P_t G_t G_r \lambda^2}{(4\pi)^3 R^4 P_n}$, and converting to decibels and substituting in

Equation 3-5, $F = S_r - SNR_{min} - 10 \log_{10} kT_0 B$, for the noise figure results in:

$$SNR_{dB} = P_{t\_dB} + G_{dBi} + 20 \log\lambda + \sigma_{dBsm} - 30 \log 4\pi - 40 \log R - S_r + SNR_{min}$$

Therefore

$$SNR_{dB} = -10 \ dBW + 19 \ dBi - 38.06 \ dB + 12.5 \ dBsm - 32.98 \ dB - 64.08 \ dB + 13.19 \ dB + 151 \ dBW/Hz$$

$$SNR_{dB} = 50.57dB \text{ for a } P_{FA} = 1 \times 10^{-6} \text{ and } P_D = 0.9.$$

Therefore, using the IPS-154 would result in a feasible solution, it must be acknowledged that losses were not taken into consideration in this exercise, meaning the actual SNR would much less. The purpose of this exercise was to obtain a theoretical estimate of the SNR the IPS-154 could achieve.

5.3   Analogue to Digital Converter

In order to obtain samples from the IPS-154 radar sensor, the system requires an analogue to digital converter (ADC). To specify the kind of ADC required, modelling the effects of the different ADC resolutions must be done. In Appendix A.2 the effects of harmonic distortion caused by the insufficient resolution of the ADC and it was found that for this application ADCs with a resolution higher than 8-bits would adequately reconstruct the signal with minimal harmonic distortions.

The IPS-154 radar sensor produces a complex-valued radar signal, meaning it has Inphase and Quadrature phase components. The proposed method of sampling this signal consists of sampling each path separately using real sampling. Thus, the sampling frequency must be twice the bandwidth of the expected radar signal [26]. These samples must be captured using two coherent channels, meaning they operate using the same clock signal in the S/H and quantization stages [32]. Once both these signals have been sampled, they can then be recombined to reconstruct the complex-valued function called the analytic signal [26].

Table 5-3 is a review of ADCs that are available in the market, the aim of this comparison is to find an ADC that must be able to sample the I/Q data coherently at an appropriate sampling rate and must have sufficient resolution to be able to reconstruct the sampled signal in post-processing without significant quantization noise.

The following are the desired specifications of the ADC for the proposed system:

- The resolution be a minimum of 8 bits.
- The ADC must have a minimum of two coherent channels.
- The SQNR must be 49.91 dB or higher.
- The sampling rate must be 100 kHz or higher.
- The system must have a dynamic range 42.14 dB or higher.

Table 5-3: Review of analogue to digital converters

Parameter	Unit	Picoscope 2206B	MAXIM Max1118	STM32L476 Discovery board	TMS320F2808 DSP	Analog Devices AD7824	Silicon Labs C8051F206
Resolution	Bit	8	8	12	12	8	12
Input channels	#	2	2	3	16	4	32
SNR @100 kHz	dB	49.92	48	65	68	46	69
SFDR @100 kHz	dB	Less than 44 dB @ ±20mV Greater 53 dB @ ±50mV or higher	66	60.24	83	42.14	80
Total Harmonic distortion	dB	-50	-69	-73	-79	N/A	-75
Internal clock	Y/N	Y	N	Y	Y	Y	Y
Max Sampling Rate	kHz	500000	100	5330	12500	100	100
Input Voltage Range	V	±0.002, ±0.005, ±01, ±0.2, ±0.5, ±1, ±2, ±5, ±10, ±20	0 - 2.7	0 - 3	0 - 3	0 – 5	0 - 3
Data Bus		USB 2.0/SPI/ I^2C	SPI	USB/SPI/USART/ I^2C/SDIO	SPI	SPI	SPI
Waveform generator	Y/N	Y	N	Y	N	N	N
Power Consumption	W	2.5	2.8	2.5	0.092	0.05	0.36
Operating temperature	°C	15 to 30	-40 to 85	-40 to 85	-40 to 85	-40 to 85	-40 to 85
Flesh Memory	kB	100000	0	512	128	0	8
RAM memory	kB	32000	0	128	36	0	1.25
Data transfer speed (Max)	Mbps	480	10	50	10	0.01	10

| Cost[11] | R | 10206.69 | 37.26 | 491.63 | 339.55 | 539.26 | 605.22 |

Table 5-3 shows multiple systems that minimum technical specifications; these systems are a combination of dedicated ADCs and DSPs with ADC sub-systems. The dedicated ADCs include the MAXIM Max1118 and the Analog devices AD7824 [44]; the dedicated DSPs with ADC sub-systems include the Silicon Labs' C8051F206, Taxes Instruments' TMS320F2808 DSP [13], the STM32L476 Discovery board and the Picoscope 2206B [45]. The primary consideration of choosing an ADC is the cost of the system and the ease of integration. The requiring external circuitry for the ADC to function is the main factor affecting ease of integration of the ADC. Unfortunately, devices which would prove challenging to integrate includes the MAXIM Max1118 ADC since it requires an external clock signal to operate, this factor eliminates this ADC as an ADC of choice [46].

The system that has the most advantages is the STM32L476 Discovery board and the Picoscope 2206B. The Picoscope can directory interface with signals smaller than 20 mV; this means that milli-volt level signals do not require pre-amplification before processing. Thus, the Picoscope has a higher effective SNR and has many data transfer protocols. The STM32L476 Discovery board is also compatible with has many transfer protocols. The two systems have many similarities, such as high data transfer speeds and waveform generation capabilities. The main advantage the STM32L476 Discovery board is the low-cost relative to the C8051F206 and Picoscope 2206b. The C8051F206 is also a compelling choice, but the low transfer speeds, a lack of waveform generator and the relatively high price disqualified it for selection [47]. The Picoscope 2206b, unfortunately, had an exorbitant price which disqualified it from solation.

Thus, the STM32L476 Discovery board is the ADC sub-system of choice for the proposed system. Since it has sampling speeds of up to 5.33 MS/s and a resolution of 12 bits, this device allows for on-site debugging as well as a dedicated digital signal processor allowing for FFT computations to be carried out on the device. This system is capable of sampling signals captured by the chosen Innosent IPS-152 radar module hence aiding in the fulfilment of Requirements 1, 2 and 3. The device cost is R 491.63, which is relatively low [48].

5.4   Radar Signal and Data Processing

In this section, the radar processing algorithms are developed. These algorithms must enable the radar to make reliable detections as well as ensure that the radar achieves the correct measurement accuracy as stated by Requirements 1, 2 and 3. The processing algorithm was prototyped using MATLAB.

The first step in constructing the radar processing algorithm was drawing a radar processing flow diagram. The first two sub-systems in Figure 5-2 consist of the acquisition of data and reconstruction of the complex signal. Figure 5-10 shows the data acquisition and reshaping flow diagram.

---

[11] The exchange rate used is R16,49

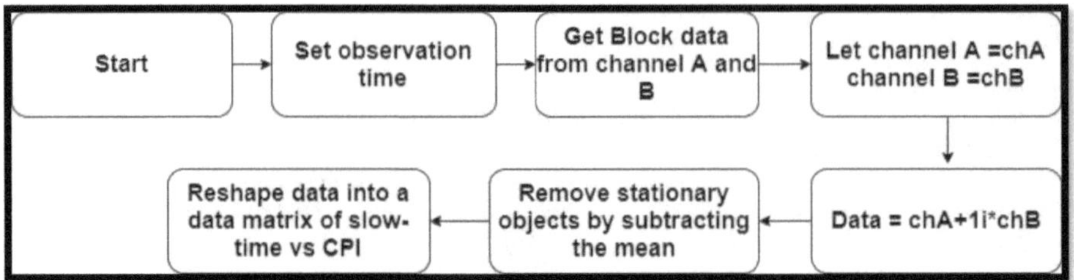

Figure 5-10: Data acquisition and reshaping.

In Figure 5-10, the data acquisition and reshaping technique were illustrated. The first step is to set the observation time of the recording; this is the coherent processing interval (CPI). The next step is obtaining the data from the device memory and into the processing workspace. Then setting up the variables and contracting a complex signal from the data is next. The mean is then removed from the complex signal in order to minimize the clutter. Lastly, the data is then reshaped into a data matrix of slow-time vs CPI.

In order to satisfy Requirement 3 and, the following considerations must be made.

Consider Equation 3.2 where $\propto = 0\ rad$ and $v\ = \pm 2km/h \approx 0.56m/s$:

$$f_{Dopp} = 2f_0 \times \frac{v}{c} \times cos(\propto)$$

$$f_{Dopp} = 2 \times 24 \times 10^9 \times \frac{(0.56m/s)}{3 \times 10^8 m/s} \times 1$$

$$f_{Dopp} = 89.6\ Hz$$

The highest Doppler frequency resulting from a car travelling at 60 km/h is about 2666.67 Hz, but since the Innocent IPS-154 sensor has a bandwidth of 50 kHz, high frequency signals may leak into the desired spectrum. In order not to complicate the external circuitry by adding an additional low pass filter, a high sampling frequency is preferred; this means that the maximum possible frequency signals must be sampled at Nyquist, which is at 100 kHz.

Sampling at 100 kHz will result in a relatively large Doppler bandwidth, which is from -50 kHz to 50 kHz. This would result in slow moving targets being missed because the dynamic range is not large enough to represent such a disparity of frequencies. Thus, an under-sampling factor of 2 must be introduced in order to decimate the samples, meaning only one sample of every ten samples will be saved. This will produce an effective sampling rate of 10 kHz [22].

Using Equation 3.13 and $f_{Dopp} = \delta_f = 89.6\ Hz, f_s = 10\ kHz$

$$\delta_f = \frac{f_s}{K}$$

$$89.6 = \frac{10 \times 10^3}{K_{min}}$$

$$\therefore K_{min} = 111.6 \sim 128\ samples$$

Where K_{min} is the minimum number of FFT points to enable a velocity resolution of $v = \pm 1.95km/h$.

In order to obtain a higher resolution, more samples must be obtained; this means a larger coherent processing interval. That results in a trade-off between latency, computational power and resolution. The unambiguous Doppler frequency range will be $\pm$ 5 kHz, resulting in the maximum unambiguously observable velocity being 125 km/h.

The 3rd sub-system calculated the signal spectrum according to Figure 5-2, this involves windowing the data, then obtaining the FFT. The window reduces the side lobes of the signal resulting in an increase in the signal to Doppler sidelobe ratio, but the main-lobe broadens, leading to a loss in frequency resolution [26]. Since the calculated resolution of this system design was higher than the required resolution stated in Requirement 1, this trade-off can be applied. The most uncomplicated window to be implemented was the hamming window, which has a maximum straddle loss of 1.68 dB. The -3 dB main lobe width (relative to a rectangular window) of 1.5 the peak sidelobe (dB relative to the peak of windowed signal) was -41.7 dB, and the maximum SNR loss relative to the rectangular window was -1.44 dB [33].

5.4.1 False Detection Avoidance Strategies

Once the spectrum has been formed, a detection decision must be made as stated by Requirement 2. That leads to the question of how false detections can be reduced or avoided. Since threshold detection is a typical detection strategy, it must be coupled with a robust system that ensures misdetections and false detections are minimised, while reliable target detections are maximised. In Section 3.7, a method of setting the detection threshold used for the detection decision was outlined by Equation 3.15; unfortunately a fixed threshold would be prone to false detections from interference as well as missed detections from increased noise present in the receiver. In the real world, the noise is not perfectly Gaussian, and the target RCS fluctuates depending on multiple parameters of the signal and target. In order to ensure the best chance of target detection, other detection strategies were investigated, and new assumptions and assertions were adopted.

Several strategies allow for more reliable target detections such as the use of clutter mapping, moving target indication and constant false alarm rate (CFAR) detectors.

Clutter mapping is a technique that allows detections of targets that have relatively low Doppler shifts. This technique is typically used for maintaining detections of targets on crossing paths, meaning targets moving orthogonal to the radar's line of sight and having zero radial velocity. This strategy can be useful for objects with relatively high RCS. Unfortunately, this strategy was not suitable in this application as the targets of interest travel radially towards the radar; hence there was sufficient separation between the target's velocity and clutter [26].

Moving target indication allows for the targets in the scene that have a velocity to be separated from the clutter. This technique works entirely in the time domain, usually using a single high pass filter. This technique is typically used in FMCW or pulse-Doppler systems. Hence, by applying an FFT operation on the data would be sufficient [26].

A CFAR detector allows changes in interference to be tracked and the detection threshold to be adjusted to maintain a constant probability of false alarm [26]. CFAR detectors are an ideal form of detection as they allow for multiple targets in the scene to be detected while minimizing false detections and missed detections.

There are many types of CFAR detectors to choose from; the choice depends on the properties of the interference, noise and the available computational power. In this study, the general assumption is that the noise and interference are homogenous.

The range of CFAR algorithms available include, cell-averaging (CA) CFAR, greatest-of CA-CFAR (GOCA-CFAR), smallest-of CA-CFAR (SOCA-CFAR), censored (CS) CFAR as well as ordered statistics (OS) CFAR and clutter map CFARs. The advantage of CA-CFAR algorithm is that it provides a dynamic threshold that adjusts to the interference and noise levels of the scene that reduces false alarms but only in homogenous environments. The CA-CFAR algorithm requires a higher signal to interference plus noise ratio (SINR) than the SNR$_{min}$ stated in Section 4.2, but it is computationally less demanding than the other CFAR stated above. The main advantage of the other CFARs (GOCA-, SOCA-, CS-, OS- and clutter map) is that they work very well on clutter boundaries for heterogeneous environments but the disadvantage is that they are more computationally expensive [26].

The commercial systems often use CA-CFAR or OS-CFAR, and it is worth comparing these algorithms to see which may be best suitable for the proposed system. The first algorithm to be described is the OS-CFAR algorithm. The input is a Doppler profile with N samples; all samples are sorted according to increasing magnitude. Which results in the ordered sequence as illustrated in Figure 5-11 [49].

$$X_1 \leq \ldots \leq X_k \leq \ldots \leq X_N \qquad\qquad 5\text{-}1$$

The statistic Z is then selected as the k-th order statistic

$$Z = X_k \qquad\qquad 5\text{-}2$$

The threshold T is a multiplication of g and K (P$_{FA}$)

$$T = Z * T_{OS}$$

Where $T_{OS} = $ K(P$_{FA}$), is a scaling factor dependent on the probability of false alarm [49]. When the threshold has been calculated, it is compared to the CUT, and then a detection decision is made. This process is illustrated in Figure 5-11.

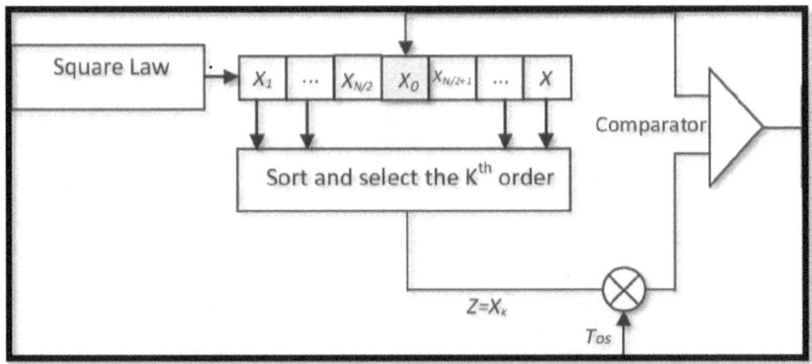

Figure 5-11: Illustration of OS-CFAR architecture [49]

The main advantage of this method is that it can discriminate between two targets in an environment with heterogeneous interference. Heterogeneous interference would include spatial, temporal variations in interference power as well as closely spaced targets returns that may bias the threshold estimates [26] [50]. In order to understand the shortfall of this algorithm, it must be compared to the CA-CFAR.

The CA-CFAR algorithm works in the following manner; first, the CFAR window resides within the data window (Doppler profile) of $z = \{z_1, z_2, \ldots, z_N\}$ and is composed of leading and lagging reference windows, guard cells (Gs), and a cell under test (CUT). Then the statistic g is obtained by calculating the mean of the f_{lead} blocks and the f_{lag} blocks. Then the two means are added and averaged. The averaged mean given as the maximum likelihood estimate.

$$\hat{g} = \hat{\sigma}^2_{\text{IIII}} = \frac{1}{N} \sum_{n=1}^{N} z_n \qquad\qquad 5\text{-}3$$

The threshold T is obtained by multiplying g and the CFAR constant. This constant is given as:

$$\alpha = N\left[P_{FA}^{-1/N} - 1\right] \qquad\qquad 5\text{-}4$$

When CUT is greater than T, then detection is declared, but when T is less than CUT, then no detection is declared [51]. This detection algorithm is visually illustrated in Figure 5-12.

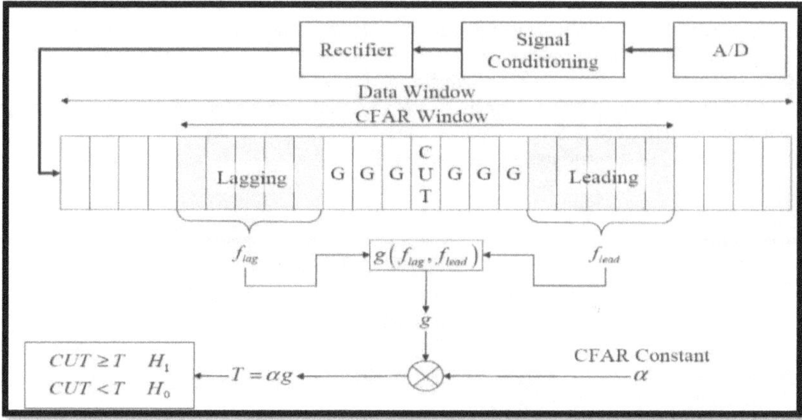

Figure 5-12: 1D CFAR architecture [26].

The main advantage of this configuration is that it exhibits optimum performance when operating in an environment with homogenous interference. The values calculated by 5.5 and 5.6 do not require complicated functions such as the gamma function to compute, which OS-CFAR is dependent on when calculating the function $K(P_{FA})$ [49].

The following is a table that compares the computational time required to compute the CA-CFAR and the OS-CFAR using fixed-point arithmetic in Q15 and Q31 format. The CMSIS-DSP library was used for the computation on a 32-bit ARM Cortex-M3 PSoC 5LP that runs at 80 MHz.

Table 5-4: The computation time for CA-, OS-CFAR [49]

fixed point arithmetic format	CFAR	
	CA	OS

| Q15 | 7ms | 13.6ms |
| Q31 | 7ms | 16ms |

The CA-CFAR has a considerable computational advantage in both formats. The environment in which the proposed system is to be placed in was assumed to have homogenous interference. Hence, the advantages offered by the OS-CFAR algorithm are not required for this application, since only one target would be observed at a time. Therefore, the CA-CFAR algorithm was chosen for this study.

There are several assertions to be made when considering a CA-CFAR detector, they are [26]:

- The interference in the reference window and the CUT is independent and identically distributed (IID).
- With a target return present in the CUT, the leading and lagging windows do not contain returns from other targets that bias the threshold estimate.
- The rectifier is square law, and thus the interference at the output is exponentially distributed.
- The mean of the interference power at the output of the rectifier is unknown and must be estimated from the samples in the reference window.
- The target is modelled as either Swerling 1 (Rayleigh voltage)
- The new SNR_{min} was 22.5 dB for the system using the CA-CFAR this was illustrated by the ROC curve in Figure 5-13.
- The P_{FA} used is 10^{-6} and has 24 reference cells.

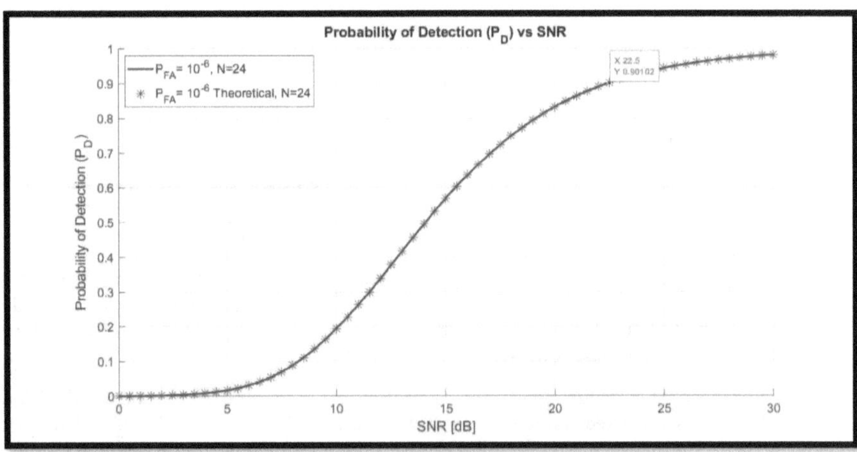

Figure 5-13: ROC curve for CA-CFAR with 24 reference cells [52]

Figure 5-14 illustrates the process of obtaining a detection vector from a CA-CFAR detector. Once detection has been made, it is stored in a vector. The detection vector contains a mixture of true detections and false detections. The CA-CFAR detector can be divided into three sub-processes, which are setting up variables, analysing the cell under test (CUT) then after iterating through all the samples.

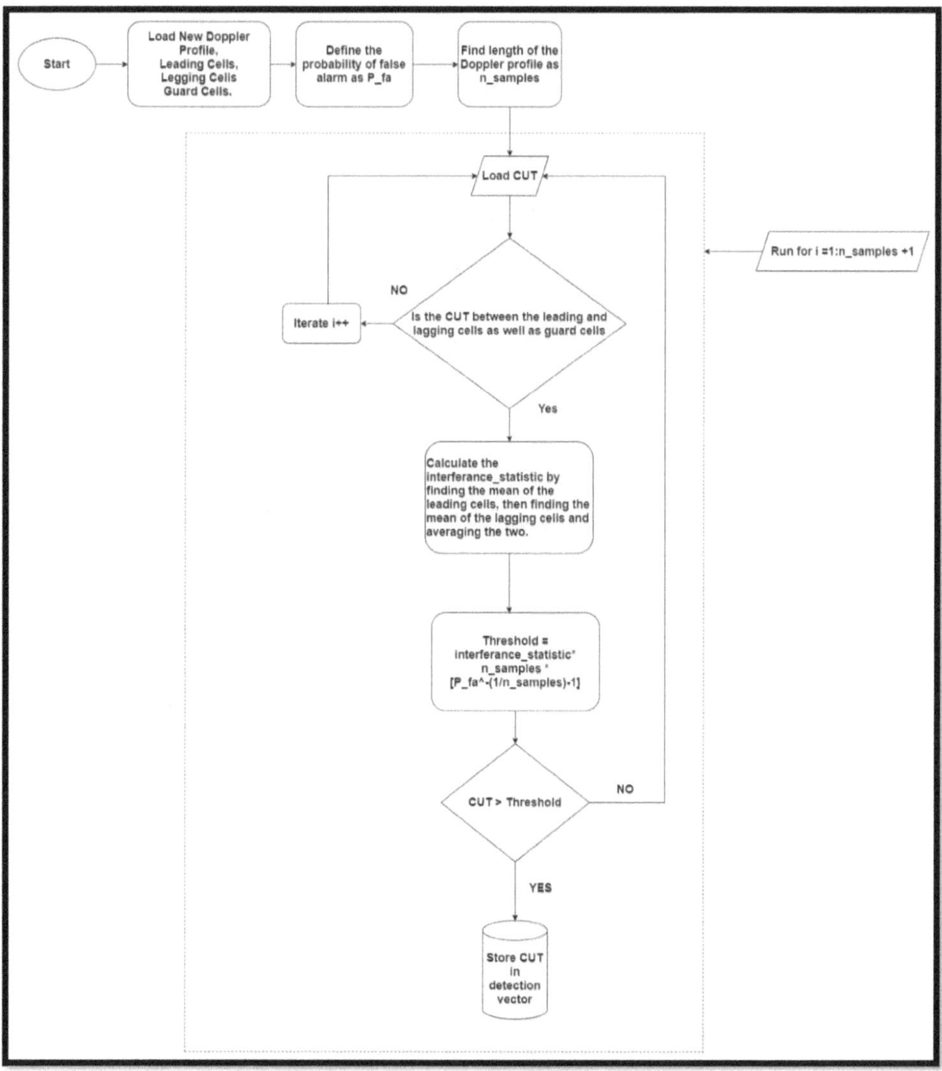

Figure 5-14: Flow diagram of CA-CFAR detector

5.5 Data Processing

This section details the data processing and analysis techniques that are typically used to ensure that the quality of detections in commercial systems.

5.5.1 Detection Association

The detection vector, as mentioned above, contains a mixture of true detections and false detections, there also exists missed detections from this vector as the probability of detection is not 100%; this presents two problems;

    i)   How can true detections that belong to a single target be associated with that target?
    ii) Missed detections.

Thus, there is a need to design an algorithm that associates every detection to the same target, to filter out false detections. There are a number of solutions that provide detection validation and the most common solutions employed in radar are detection association and tracking. Target tracking typically consists of two parts, track filtering and measurement-to-track data association. Track filtering consists of estimating the trajectory of a track (i.e., velocity, position and acceleration) using measurements associated with a track, e.g. range elevation and bearing. Measurement-to-track data association or data association is a process of assigning a measurement to an already existing track or detection of a newly formed track associated with a new target or a false signal [26].

There are a number of tracking algorithms and data association algorithms used in industry, tracking algorithms such as Kalman filtering, alpha-beta filtering and the interacting multiple model, as well as data association algorithms such as the statistical nearest neighbour, strongest neighbour and the probabilistic data association filter [53]. These advanced topics were beyond the scope of this study since it can be assumed that only one car would exist in the beam at a time there would be no need to distinguish two targets from each other. The dimensions this proposed system can measure was Doppler and time; tracking algorithms typically work with targets in a 3D space. Thus, the CA-CFAR is the only means for false detection avoidance. The proposed system does not make use of any association algorithms custom or otherwise.

Figure 5-15 illustrates the flow diagram of the above-mentioned process.

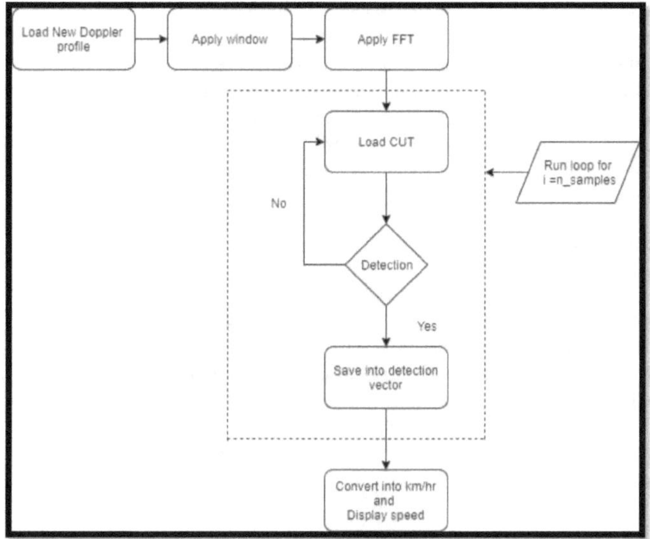

Figure 5-15: Radar signal processing algorithm.

The next step in developing the signal processing algorithm is choosing a programming language/framework for the execution of the program. The primary considerations for a programming language/framework to be chosen are, the availability of signal processing libraries, ease of use, access to the IDE and the last consideration is the compatibility with the DSP.

The main programming languages/framework to be considered in this study are:

- MATLAB code

- Python 2.7/3

- Julia

- C/C++

- Java

There many more programming frameworks available in the industry, but this study will only focus on the above-mentioned frameworks. These chosen frameworks are the most common frameworks used in the CSIR radar research group. Should it be required, support from senior researchers is available.

5.5.2 MATLAB/M script

MATLAB Matrix laboratory (MATLAB) is defined as a multi-paradigm numerical computing environment and proprietary programming language; it is the most common programming framework used in engineering research institutions and universities. MATLAB has a number of signal processing toolboxes, including specialized radar signal processing toolbox. MATLAB has several advantages; this includes an easy to use IDE, and simple to learn syntax. MATLAB support is ubiquitous; it also has highly optimized algorithms for matrix calculations, including FFTs. The main disadvantages with MATLAB are that its algorithms are proprietary, which means that they are not published to see if they are implemented correctly. It is also costly; all toolboxes are sold separately; however, student trails are available. The last disadvantage is that the latest MATLAB is computationally heavy, requiring a very powerful platform to run it. It requires a 64bit machine with a minimum of 2.9 GB of hard drive space as well as 4 GB of RAM, however standalone MATLAB functions are deployable on 32-bit machines with an ARM-based processor [54].

5.5.3 Python 2.7/3

Python is an open-source high-level, general-purpose programming language. Python is used in web development, games development as well as in science and engineering. It has a large community in both industry and academia. This framework has the advantage that it is easy to read and learn. The code developed in this framework can work on most devices, including 32-bit devices. This framework has extensive scientific libraries which support signal processing operations such as highly optimized FFTs. The significant disadvantages of Python are that it is an interpreted language meaning that it is not optimized for use on dedicated hardware and some programs may be slow to execute [54].

5.5.4 Julia

Julia programming language is an open-source, high level, high-performance dynamic programming framework. It was specifically built for numerical analysis and computational science. The language is similar in syntax with Python and MATLAB. The language is highly optimized for operations such as FFTs and other signal processing algorithms. It is a reasonably recent framework, only being launched in 2012, this means it does not have as much a large industry presence and is mainly used in research institutions. The Julia community is growing, but it is not as large as the Python and MATLAB communities. The platform also runs on 32-bit machines [55].

5.5.5 C/C++

The C/C++ programming framework is a general-purpose, high-level language, which is generally associated with speed and efficiency. This programming language is very different in syntax to the previously discussed programming languages; it relatively complicated. It is a highly established language and has a large community in both industry and academia. It has several signal processing algorithms which are highly optimised and would give near real-time performance in low powered devices [56]. While C/C++ programs may run on both 32 and 64-bit machines, developing this program requires very bulky IDEs such as Microsoft Visual Studio, this requires a 2.6 GHz processor, 4 GB of RAM and at least 10GB of hard disk storage.

5.5.6 Java

Java is a high-level programming language; it was designed to have the look and feel of the C++ programming language but is simpler to use and enforces an object-orientated programming model. Its applications can run on a single computer or be used in distributed among servers. Java applications can run on 32-bit machines and higher through Java Standard Edition (SE) Embedded. This framework has a large community and has highly optimised signal processing algorithms. The main disadvantage of JAVA is the ease of use and a complex syntax. The programs developed using java require the latest updates of JRE which help the programs run of the java virtual machine; this limits the speed of the programs [57].

The frameworks outlined in this section are all deployable on both

Table 5-5: Comparison of programming frameworks

Conditions	Programming framework/language				
	MATLAB	Python	Julia	C/C++	Java
Signal Processing Libraries	Yes	Yes	Yes	Yes	Yes
Ease of use (Syntax)	Yes	Yes	Yes	No	No
Ease of use (IDE)	Yes	Yes	No	No	No
Affordability (cost)	No	Yes	Yes	Yes	Yes
Support	Yes	Yes	No	Yes	Yes
Compatibility with DSP (32 bit-64 bit)	Yes	Yes	Yes	Yes	Yes

From Table 5-5, it can be seen that the programming framework that meets all stipulated conditions is the Python programming framework. Thus, the code shall be written using Python.

5.6   Digital Signal Processor

In order to process the data in near real-time to satisfy Requirements 1, 2 and 3; it is critical to review DSPs that are available on the market. This comparison aims to find a DSP that will be able to provide near real-time speed estimates. Consider a car travelling at 60 km/h, that has been detected at 40 m from the radar. The radar must provide a speed estimate within half a second from the detection; this means that the motorist will have four measurement readings in the 40 m span if they do not change their speed.

- The maximum computational and display time must be less than 500 ms.

Table 5-4 in Section 5.4.1 provides estimates of computational times of the CA-CFAR using Q31 and Q15 fixed-point arithmetic on a 32-bit ARM Cortex-M3 PSoC 5LP that runs at 80 MHz [58]. The CA-CFAR algorithm takes 7 ms for both Q formats when using 1024 samples, where 256 are real samples from the ADC, and the 768 samples are zero-padding. The entire processes, including sampling, finding the complex-FFT, and obtaining the speed estimate, took 27.102 ms.

Table 5-6 is a review of standard multi-purpose DSPs. A comparison of different specifications of these devices was made in order to see which would be most suitable to be used in the proposed system. The primary considerations are in relation to how quickly these DSPs compute standard algorithms such as real and complex FFTs; This would be the measured using figures from Table 3-2 in Section 3.6.1 and the processor core speed. The processor core speed is measured in millions of instructions per second (MIPS) for central processing units (CPUs) and millions of floating-point operations per seconds (MFLOPS) for graphical processing units (GPUs) and floating-point DSPs, since most DSPs only operate with fixed-point values and not floating-point values, this study focuses on fixed-point operations using the Q format in order to represent fractions [59].

Table 5-6: Review of Digital Signal Processors

		Model							
Parameter	Units	Raspberry Pi 3 Model B+	Orange Pi 3	Beagle bone Green	Intel Edison	ADSP-21060LC	STM32L476 Discovery board	TMS32 0F2808	Arduino Uno
CPU Clock speed	MHz	1200	1800	1000	400	400	80	100	16
Processor core speed	MIPS	2441	1800	2000	615	40	100	100	16
RAM	MB	1000	2000	512	1000	0.5	0.128	0.036	0.008
Flash Memory	MB	Expendable up to 32000	8000	4000	4000	0	0.512	0.128	0.256
GPIOs	pins	40	26	46	28	10	114	16	20
Network Connectivity	Bluetooth	4.1	N/A	4.1	4	N/A	N/A	N/A	N/A
	Ethernet	1000 Mb	1000 Mb	N/A	N/A	N/A	N/A	N/A	N/A
	WIFI	2.4 GHz	2.4 GHz	2.4 GHz	2.4 GHz	N/A	N/A	N/A	N/A
External Interfaces	USB	2.0	3.0	2.0	2.0	N/A	2.0	N/A	2.0
Weight	g	45	75	81	70	14	60	16	25
Power	W	12.75	12.75	12.75	0.32W	0.0095	0.032	0.092	3.5
Price	R	598.15	1178	880	2985.00	3303.46	491.63	339.55	270

In order to satisfy Requirement 3, which states that the system shall be accurate to $\pm$ 2km/h, this would require 128 FFT points corresponding to 1792 real multiplication instructions. Therefore, to obtain the time it takes to compute this 128-point FFT, the processor speed was divided by the number of instructions. The result is Table 5-7.

Table 5-7: Comparison between the computation times for the DSPs in Table 5-6.

Parameter	Units	Raspberry Pi 3 Model B+	Orange Pi 3	Beagle bone Green	Intel Edison	ADSP-21060LC	STM32L476 Discovery board	TMS320F2808	Arduino Uno
CPU Clock speed	MHz	1200	1800	1000	400	400	80	100	16

| Processor core speed | MIPS | 2441 | 1800 | 2000 | 615 | 40 | 100 | 100 | 16 |
| Computation time for 128-point FFT | µs | 0.73 | 0.9956 | 0.896 | 2.91 | 44.8 | 17.92 | 17.92 | 112 |

It can be observed by the comparison in Table 5-7 that the Raspberry Pi offers higher processing speeds compared to dedicated DSPs such as ADSP-21060LC, TI's TMS320F2808 and STM32L476 [60]. The cost is also lower than its immediate competitor which is the Orange PI 3. The power considerations are apparent since it uses 12.75W peak power.

This board also has 1 GB of low-power double data rate (LPDDR2) static random-access memory (SRAM). The Raspberry Pi offers expandable flash memory of up to 32 GB, of which 29.8 GB is usable memory. The Raspbian Jessie operating system requires 4.3 GB; only 25.5 GB may be used to save more than ten days' worth data hence satisfying the 7th requirement. This system also has WF-FI and supports 1 Gbps but is limited to ~300 Mbps Ethernet due to USB 2.0 limitations; these specifications allow for internet connectivity for remote data retrieval and analysis. The Raspberry Pi shall be used as the DSP of choice in this design [60].

In order to obtain the total computation time for one velocity estimate, the time taken to execute the following sub-process must be added together. These processes include the ADC sampling frequency, ADC data transfer rates, digital signal processing speed and the SPI data transfer rate. The SPI data transfer rate is the time taken for the final velocity estimate data to be sent to the two digits 7-segment display. The systems computation specifications are presented in Table 5-8.

Table 5-8: The system computation variables.

ADC sampling frequency	ADC transfer rates	DSP computation speed	SPI transfer rate
10 kHz	50Mbps	2441 MIPS	10 Mbps

It takes 128 samples (1024 bits) to obtain a velocity estimate with a doppler resolution of ± 2 km/h, this includes 1792 FFT multiplication instructions, and the 1152 operations of the CA-CFAR; since it performs nine operations on every sample. The velocity estimate result is a binary coded decimal (BCD) that is 8 bits long since two 7-segment displays are used. Thus, the total theoretical compute time were obtained by adding separate computational times shown in Table 5-9 for CPIs of 0.0128 s, 0.0256 s and 0.0512 s .

Table 5-9: Velocity estimate computation time.

ADC sampling time	ADC transfer time	FFT computation time	CA-CFAR computation time	Display time	Total computation time
12.8 ms	0.02048 ms	0.73 µs	0.47 µs	800 ns	12.82 ms
25.6 ms	0.0401 ms	1.68 µs	0.94 µs	800 ns	25.64 ms
51.2 ms	0.0819 ms	3.78 µs	1.68 µs	800 ns	51.29 ms

The theoretical compute times are all less than the measurement update time, this means that real-time updates are possible; this result aligns with Requirements 1,2 and 3.

5.7   Display

This section seeks to satisfy the condition that the driver must be able to see their speeds up to 40 m from the radar while travelling at a peak speed of 60 km/h, this is the 4th Requirement. The display must be large enough and bright enough to be seen in the day and night-time.

The luminance flux, "which represents the light power of a source as perceived by the human eye" [61] , of the display is a critical variable in the power and cost calculations. Consider Table 5-10, which shows the typical values for the luminance of different conditions.

Table 5-10: Illumination of different environments [61].

Illumination condition	Illuminance
Full moon	1 lux
Street lighting	10 lux
Home lighting	30 to 300 lux
Office desk lighting	100 to 1 000 lux
Surgery lighting	10 000 lux
Direct sunlight	100 000 lux

Where illumination is the luminous flux incident per unit area, which is measured in lux $= \text{lm/m}^2$ $= \text{cd/m}^2$, thus the display must be visible at 100 000 lux. Since light is additive, the light display must be a single frequency/colour to be visible during the daytime [61]. The LED brightness is measured in milli-candela (mcd), to convert from mcd to lux the distance from the LED source must be considered.

$$E_v = \frac{I_v}{d}$$

5-5

Where

E_v: Illuminance in lux

I_v : Luminous intensity in cd

d: distance form LED in meters

The following is the types of displays that are typical for this application.

Table 5-11: Review of LED displays.

Parameter	Unit	LYI LIYI-NB-R	Xuancai	Cebek CD-29 7-Segment Super Bright Red BCD LED Display Module	TUOXING TD0440	IR5013RI	Bluewin-led
				Manufacturer			
Colour	R/G/Y/B/W	R	R	R	G	R	R
Size (L x W x H))	cm	41 x 5 x 25	40 x 6.2 x 21	29 x 1.5 x 36	53 x 6 x 20	66 x 3.3 x9 .7	98 x 6 x 38
Brightness at 40 m	lux	0.003125	0.0008125	0.001875	0.0002813	0.0014375	0.00375
Power consumption	W	45	18	16.8	8	25	80
Supply voltage	V AC	220	100-240		100-240	220	220
AC/DC	V DC			21			
weight	Kg	1.5	1.5	0.22	2.6	3.2	6
Cost	R	1022.93	583.80	2198.16	2 013.54	5306.68	2 445.80

The main objective of this exercise is to find a display that is bright enough to be seen from 40 m. Essential considerations in making this selection are energy usage, the weight of the system and cost. A trade-off exists between the power consumption and the cost since bright displays are more energy consuming and costly. The Bluewin-Led display is the brightest at 40 m with an illuminance of 0.0375 lux, but the power consumption is relatively high at 80 W. This system also requires 220 V AC which means an inverter must also be bought in order to adequately supply it with power, as with most systems reviewed; this will increase costs considerably. The best option is the Cebek CD-29 7-segment display module; this system is powered by 21 V DC supply at 16.8 W. It has the third-highest brightness from the reviewed systems and is relatively lightweight at 0.22 kg. It does, however, have a relatively high cost at R2198.16, but this cost is offset by the energy and efficiency considerations since this system does not require a DC to AC inverter.

5.8 Power System

The power budget and system cost shall be outlined in this section; this is to satisfy Requirement 6. The power requirements system components are presented in Table 5-12.

Table 5-12: Power budget of proposed system

Component	Power usage (W)
IPS-154 Radar sensor	0.25
STM32L476 Discovery board	0.03
Cebek CD-29 7-Segment Super Bright Red BCD LED Display Module	16
Raspberry Pi model 3 B+	12.75
Total Power	29.03

Since the Raspberry Pi can provide up to 5 W of power at 5 V DC, the ADC and radar module shall be powered by it. Thus, the effective power consumption is 29 W. This system requires a battery and a solar system solution. The battery allows the system to be self-sustaining, and the solar system charges the battery for continuous use. The solar system requires a solar charge controller to work and allow the system to charge fully while the sun is out, which is typically 12 hours.

There needs to be a water-resistant enclosure as well as a 1.5 m steel pole to carry the combined system.

Consider a battery with a 12V x Ah DC power rating that is required to last 12 hours. The rating should be calculated using the following:

The required power is approximately 30 W

$$Time\ on = \frac{Power\ rating\ battery}{Power\ required} \qquad \text{5-6}$$

$$12\,h = \frac{(12\,V \times x.\, \wr\, h)}{30\,W}$$

$$x = 30\ Amp.\,hours$$

A review of batteries that can fulfil this requirement was also made, and Table 5-13 is a summary of the reviewed components.

Table 5-13: Review of deep cycle battery sources

Quantity	Unit	CSB EVX12300 Deep Cycle AGM VRLA Traction Sealed Lead Acid Battery	DEEP CYCLE GEL SOLAR BATTERY 12V - Gamistar	GOLDSHINE DEEP CYCLE GEL SOLAR BATTERY 12V	Victron energy AGM super cycle battery	Victron energy Deep Cycle GEL
Capacity	A.h	30	50	50	38	66

		12	12	12	12	12
Voltage	V DC	12	12	12	12	12
Time on @ rated DoD	hours	3.6	12	6	4.56	7.92
Size	L x W x H (cm)	16.6 x 12.5 x 17.5	32 x 16 x 17	33 x 17 x 17	19.7 x 16.5 x 17	25.8x 16.6 x23.5
Weight (approx.)	kg	9.25	18	19	12.5	24
Operating temp	°C	25	25	20-30	20 to 30	25
Solar charging	Yes/No	Yes	Yes	Yes	Yes	Yes
Battery life	Years	8	10	8	7 -10	12
	Recharge cycles	1800	1200	1800	1500	1800
	% Depth of Discharge	30	60	30	30	30
Cost	R	1529.95	1349	1349	2280	2799

The energy solution required must provide 30 W for 12 hours; the systems reviewed consisted of absorbent glass mat (AGM) and lead-acid gel batteries that can be recharged using solar power. The selection of the optimal battery system is a trade-off between the system on time, the number of rated recharge cycles and cost. Careful attention must be given to the weight of the system as batteries are a significant contributor to the system weight.

The minimum rated battery in this review is a 12 V 30 Ah AGM system that has a depth of discharge (DoD) of 30%; which means that this system can only provide 3.6 hours of on-time at 30 W because of the imposed DoD. Most of these batteries have a DoD of 30% when aiming for recharge cycles higher than a thousand. However, these systems may also be used at 50% DoD, but this significantly reduces the number of recharge cycles they can undergo; for instance, the Victron energy Deep Cycle 12 V 66 Ah gel battery can only withstand 750 recharge cycles at 50% DoD while the Victron energy 12 V 38 Ah AGM supercycle battery lasts for 600 recharge cycles at 50% DoD. Low recharge cycles are the general trend of AGM batteries as opposed to gel batteries; thus, the proposed battery should be a gel battery [62].

Unfortunately, the system requires 12 hours of on-time at 30W this means that the system that only the 12 V 50 Ah deep-cycle gel solar battery manufactured by Gamistar is the only solution that can meet this criterion as it has a DoD of 60% at 1200 cycles; this means it can provide 12 hours of on time as well as last 1200 recharge cycles. The cost of the system is also modest as it costs R1349 [63].

The system requires two deep-cycle solar gel batteries since the system operates during the day, and that is the only time the solar panel can provide power. A solar panel solution must adopt a staggered approach to recharging the batteries; this means the system will use one battery for its energy needs for 12 hours and the next day the full battery will be used while the battery at 60% DoD recharges. Thus, the total cost of the battery solution is R2698 and weighs 36 kg.

Efficiency considerations

The system requires 30 W of power to operate under normal conditions; the proposed battery can provide 30 W of power for 12 hours at the recommended 60% DoD. The solar panel solution must be able to provide 30 W of power when there is an overcast day. Typically, solar panels can produce 10 -25% of their rated capacity on cloudy days. Since the solar panel must produce 30 W of power on an overcast day (10% of rated power), therefore the total rated power of the solar panel must be 300 W [64]. Table 5-14 is a review of solar panels available on the market.

Table 5-14: Review of 300 W solar panels

Parameter	Unit	Manufacturer				
		CanadianSolar BiKu Module	The Sun Pays 300W mono Solar panel	Renewsys	Fivestar	ARTsolar
Rated power	W	300	300	330	300	300
Efficiency	%	17.89	18.3	17.08	17	18.5
Operating voltage (Vmp)	V	32.5	32.6	37.62	36.6	32.6
Operating current (Imp)	A	9.24	9.19	8.78	8.2	9.21
Operating temperature	°C	-40 to 85	-40 to 85	-40 to 85	-40 to 85	-40 to 85
Dimensions (L x W x H)	cm	169 x 99.2 x 0.58	164 x 99 x 3.5	195.7 x 98.7 x 4	196 x 99.2 x 4	164 x 99.2 x 4
Weight	kg	24.3	18.2	21.5	25	18
Cost	R	1665	2195	2160	2993	1650

Table 5-14 lists a select number of 300 W solar panels; the choice of solar panels is influenced by weight, cost and efficiency. Since most of the parameters of these solar panels are similar, the most outstanding system from this selection is the ARTsolar 300W solar panels. They have an 18.5% efficiency; they weigh 18 kg and cost R1650.

The ARTsolar 300 W solar panels must be paired with a capable solar charge controller that can handle two 12 V 50 Ah deep-cycle gel solar batteries. Table 5-15 is a review of available solar controllers able to handle the specifications laid out above.

Table 5-15: Review of solar charge controllers

Parameters	Units	Manufacturer				
		PWM SKU: 11324-40A	SolaMr 100A MPPT Solar Charge Controller	Solar Charge Controller Model: LD2440CWM	GAmister 40A Solar charge controller	Ashcom Solar Charge Controller 12/24V – 50A
Type	MPPT /PWM	PWM	MPTT	PWM	PWM	MPTT
Conversion rate	%	80	99	80	80	99

Rated voltage	V	12/24	12/24	12/24	12/24	12/24
USB output	Yes/No	Yes	Yes	No	yes	No
	V/A	5/2	5/2		5/2	
Rated current	A	40	40	40	40	50
Max solar input voltage	V	50	50	36	40	40
Max solar input current	A	50	50	30	40	50
Weight	g	174	350	180	173	320
Cost	R	356.56	1149	699	762	899

It can be observed from Table 5-15 that there are two types of solar charge controllers, the pulse width modulation (PWM) scheme and the maximum power point tracking (MPPT) scheme. The main difference between the two is that MPPT solar charge controllers are more efficient than PWM solar charge controllers; this also has a bearing on cost as MPPT controllers are relatively more expensive. Since the solar panels are only 300 W, the PWM scheme is more suited as the inefficacies have already been compensated. This chosen system is the PWM SKU: 11324-40A solar charge controller; this system can provide USB power at 5V DC 2A and is also much cheaper than the other solar charge controllers at only R356.56.

The total weight of the electronics of the system is 61kg and the internal dimensions of the system must be able to fit two 12 V 50 Ah deep-cycle gel solar battery manufactured by Gamistar, they have dimensions 32 x 16 x 17 cm. the next section the design of the enclosure and steel pole will be made. The 6th requirement states that the system must have a combined weight of 48 kg, as typical systems weigh this much; the reviewed systems has smaller batteries and solar panels. The designed system has twice the battery and solar capacity.

5.9   Enclosure, Pole, Electrical Connection and Cost

The design of the steel pole must be able to with stand 61 kg of weight and the enclosure must allow for two deep-cycle gel solar battery manufactured by Gamistar with dimensions 32 x 16 x 17 cm.

Consider the following design:

Rectangular enclosure with dimensions (L x W x H) 50 x 20 x 60 cm, which is 3.5 the hight of the batteries. This means the batteries can be stacked on top of each other and still low for 26 cm of clearance.

The steel pole must support a minimum 61 kg.

The gravitational force (W) of the box:

$$W = mg \qquad\qquad 5\text{-}7$$

Where

g: The gravitational acceleration of 9.8 meters per second

Using Equation 5-7,

$$W= 61 \text{ kg} \times 9.81 \text{ m/s}^2$$

$$= 598.41 \text{ N}$$

$\therefore$ The pole must be able to carry a load of 598.41 N. Checking whether the pole is adequate to carry the proposed load. Assume the outside diameter of pole is 114.3 mm and the thickness of the steel of the pole is 2.5 mm.

Section classification:

$$\frac{d}{t} = \frac{114.3}{2.5} = 45.72$$

$\therefore$ Limiting $\frac{d}{t} = \frac{23\,000}{f_y} = \frac{23\,000}{350} = 65.7 > 45.72$

Where

f_y: Specified minimum yield stress

$\therefore$ Cross-Section is not Class 4 (slender)

Using the Slenderness ratio [65]:

$$\left(\frac{kL}{r}\right)_x = \left(\frac{kL}{r}\right)_y = \frac{3000}{39.5} = 75.94 < 200 \text{ (limiting factor)} \qquad 5\text{-}8$$

Where

kL : Effective length in meters

r: Radius of gyration of a member about its weak axis in meters

Compressive resistance [65], [66]:

$$f_e = \frac{\pi^2 E}{(kL/r)^2} = \frac{\pi^2 \times 200 \times 10^3}{(75.95)^2} = 342.25 \text{ MPa} \qquad 5\text{-}9$$

Where

E: Elastic modulus of Steel in MPa

$$\lambda = \sqrt{\frac{f_y}{f_e}} = \sqrt{\frac{350}{342.29}} = 1.011$$

$$C_r = \varphi A f_y (1 + \lambda^{2n})^{\frac{-1}{n}} \qquad 5\text{-}10$$

$$C_r = 0.9 \times 0.898 \times 10^3 \times 350 \times 10^{-3} (1 + 1.011^{2.68})^{\frac{-1}{1.34}}$$

$$C_r = 163.03 \text{ kN} > 598.41 \text{ N}$$

Where

C_r: Compressive resistance (Maximum weight the structure can carry)

A: cross-Sectional area of bolt, based on normal diameter in meters φ: Resistance factor for structural steel

λ: Non-dimensional slenderness ratio in column formula

n: Number of shear connectors required between point of maximum positive bending moment and adjacent point of zero moment

$$m = \frac{W}{g} = \frac{16302.82}{9.81} = 16618 \text{ kg}$$

The steel pole can carry a mass of 16 618 kg; which means that the 6th requirement must be amended to reflect the dimensions of the enclosure and ability of the steel pole to withstand the system weight.

The resultant design is shown in the technical drawing in Figure 5-16

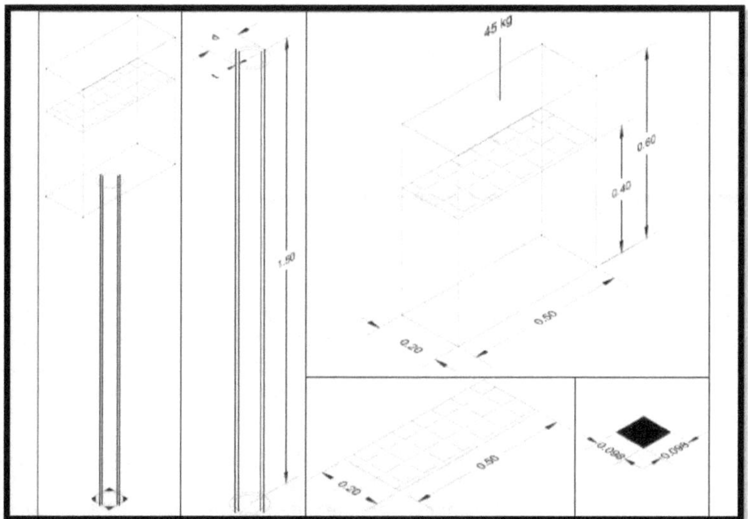

Figure 5-16: Resultant design of enclosure with a steel pole

Figure 5-16 shows that the enclosure has two chambers, the lower chamber can accommodate two deep-cycle gel solar battery manufactured by Gamistar stacked on top of each other. The upper chamber will house the solar controller, the DSP and ADC. There enough space on both chambers to house all the components specified, thus the amended 6th requirement has been fulfilled.

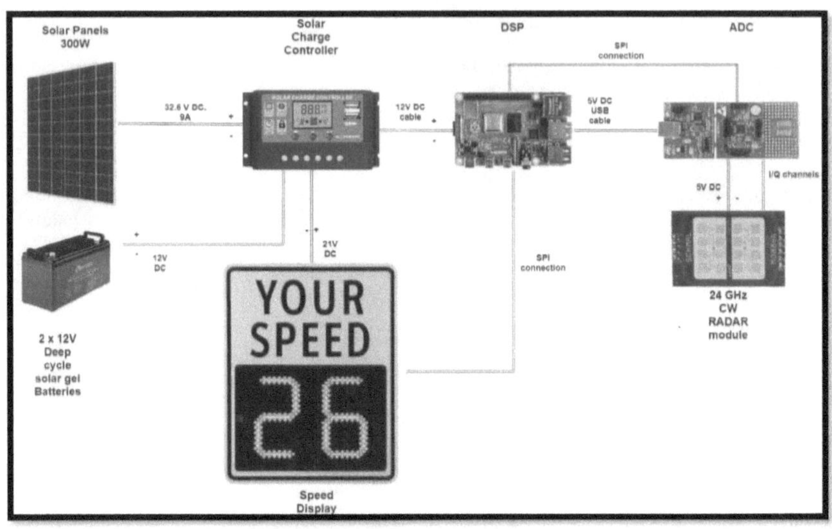

Figure 5-17: Electrically connected schematic of proposed system.

Figure 5-17 shows an electrically connected schematic of the proposed system. The power system consists of a 300 W solar panel and two deep cycle 12 V batteries, the voltage provided by the solar panel and the battery is 36 V DC, and the current is 9 A. These components are connected to a solar charge controller. The controller provides power to the speed display as well as the Raspberry Pi. The solar charge controller is also responsible for charging the battery when it is at about 60% DoD.

The Raspberry Pi obtains power through a 12 V DC cable. The Raspberry Pi then powers the discovery board through a 5 V DC USB cable. The Discovery board powers the IPS-154 radar sensor through its 5 V pins. The I/Q channels from the radar module send the baseband analogue signal to the ADC which then sends the digitized signal through an SPI connection to the DSP; then another SPI connection transfers the estimated speed from the DSP and the speed display.

Table 5-16: Bill of materials

Component	Cost (R)
IPS-154 Radar sensor	850
STM32L476 Discovery board	491.63
Cebek CD-29 7-Segment	2198.16
Raspberry Pi model 3 B+	598.15
Solar Panel (300W)	1650
Battery	2698

Solar charge controller (40A)	356.56
Waterproof enclosure	1506.5
1.5m pole	499
Total cost[12]	10848

A custom-made waterproof steel enclosure that is 50 x 20 x 60 cm costs R1506.5 which is IP65 water and dust resistant. This enclosure shall house all the components except the radar module, display and solar system. The total cost of the system components is R10848, and the total power consumption is 30 W. The labour of a skilled technician to build such a system is R450/hour[13], the estimated time of assembly of the system is 3.5 hours[14]. Therefore, the total labour cost is R1575; this results in R12423 as the total cost of the proposed system. This final figure satisfies last Requirement 7 of the non-technical requirements.

5.10 Data collection system

This chapter details the design and specification of low-cost traffic calming radar, demonstrating the feasibility of such a system. The proposed system would cost a total of R10465.88, which is well above the budget of this project of R5000. There is a need to obtain real data for the sensor to prove the viability of the proposed system. This technology demonstrator must validate the proposed system in key performance areas, such as vehicle velocity estimation accuracy, as well as detection range accuracy.

The proposed system's performance requirements have been demonstrated in the computation time calculations. Since the Raspberry pi has benchmark scores from [60], it is a reasonable assumption that this system can handle the real-time velocity estimation calculations given the proposed system's theoretical processing speeds which are listed in Table 5-9. Unfortunately, the radar sensor proposed for this study must be procured as there is no physical way to approximate the data collected from this device.

In order for Requirement 1,2 and 3 to be satisfied, the data obtained using the radar sensor must be processed and analysed. The data-collection system is tasked with capturing the data; this data will then be processed off-site to determine the feasibility of the proposed system.

The radar data must be analysed using the algorithms proposed in Section 5.4; thus a prototyping language may be used to efficiently determine the validity of the sensor data and thus satisfying Requirements 1, 2 and 3. MATLAB is a useful language often used to quickly and accurately implement signal processing algorithms. Since there exist student trails of the software, the proposed algorithms may be coded using MATLAB, and the results can then analysed to validate whether the data obtained from the sensor is sufficient to satisfy Requirements 1, 2 and 3. It was stated in Section 1.7 that the proposed system does not have to be built as this study is only a feasibility study of a traffic calming solution for the CSIR.

The data-collection system consists of the chosen radar sensor which is the Innosent IPS-154 CW radar module, the Picoscope 2206B portable oscilloscope and a Samsung Series 7 Chronos Laptop. The Picoscope 2206B portable oscilloscope allows signals that are ±20 mV and has 8 bits resolution; thus, no amplifier is required to increase the signal levels. This system can also produce signals within the Doppler bandwidth of interest through its arbitrary waveform

---

[12] Total cost at the time this report was written, prices might be subject to change.

[13] The quoted price of Technician working at the CSIR Labs

[14] The estimated time is given by Technician, estimated by the complexity of the system.

generator; this is very useful as preliminary tests may be conducted using this system. The real-time FFT function built into the Picoscope 2206b also allows for the analysis of harmonic distortion present in the system and thus, aiding in the algorithm development of a robust CA-CAFR.

5.11 Summary

This chapter outlines the design of a traffic calming solution. The validation of design decisions was achieved through relevant design equations and calculations. A system that meets the user requirements was then outlined, and the different sections and sub-sections showed how each sub-system operated and intended to satisfy the relevant technical requirements.

An amendment to the 6th requirement was proposed as the system design weight was found to be 61 kg instead of 45 kg; this was not an issue as the steel pole designed has a very high compressive resistance relative the combined weight of the system components. A bill of materials was then compiled using quotations from different suppliers, and the total cost was below the R20k budget, thus satisfying the 7th requirement. Section 5.10, then outlined why a data collection solution must be built in order to demonstrate the validity of the data form the chosen radar sensor and how the data will be analysed using signal processing algorithms implemented in MATLAB.

Chapter 6

System Integration and Testing

In this chapter, the data-collection system was integrated and tested to confirm that both the signal level and signal processing requirements are satisfied. Section 4.4 details all the experiments that are required for system validation. Before the system can be tested as a whole, each sub-system first needs to be tested to determine the correct operation. Once this is established, the experiments detailed in Section 4.4 may be conducted.

The signal processing and data analysis was implemented on a Samsung Series 7 Chronos Laptop. The insight gained in this chapter may be used to accelerate prototype development and enable design modifications that will lead to improved system design.

6.1   Sub-system Integration

In this section, the details of the integration of the sub-systems that make up this system are provided. The fully integrated experimental system is shown in Figure 6-1.

Figure 6-1: Fully integrated experimental radar system.

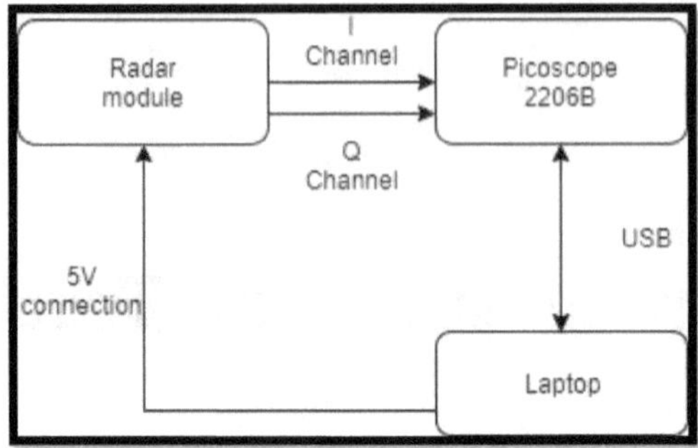

Figure 6-2: Connections of the different sub-systems

The radar module and the Picoscope have separate connections to the laptop. The ADC uses the same USB connection for both power and data transfer. The radar module is connected to the laptop through a 5V USB A connector for power only. The radar module only has an analogue connection to the Picoscope through the I and Q channels.

6.2 Sub-system Testing

In this section, the separate sub-systems were tested to confirm regular operation and to benchmark each sub-system against the design parameters. Unfortunately, the power system and the display system were not tested as stated in the limitations of this project. These two sub-systems shall be assumed to be functioning as intended by the design specifications. The experimental system intends to demonstrate the feasibility of using the chosen radar transceiver module as a traffic calming solution as the data obtained from this module is the primary determinant of this premise.

6.2.1 Radar signal processing

this section, the radar processing algorithms were tested to ensure the correct functionality of each component of the radar system software. The radar signal processing chain was shown in Figure 5-2, and the processing chain begins with the acquisition of data, then the reconstruction of the data into a complex signal. The spectrum of then signal follows, together with the filtering of unwanted signals and noise. The last step pertains to the detection of targets in the signal. The detailed signal processing algorithm is presented in Section 5.5.1 Figure 5-15. All the steps shown in this flow diagram was probed to see if the results of each step were correct; this was done using a point scatterer model data that represented a car approaching a radar at 60 km/h.

The first step pertains to the acquisition of time-domain signal samples for one CPI of 0.0128 s, as is shown in Figure 6-3.

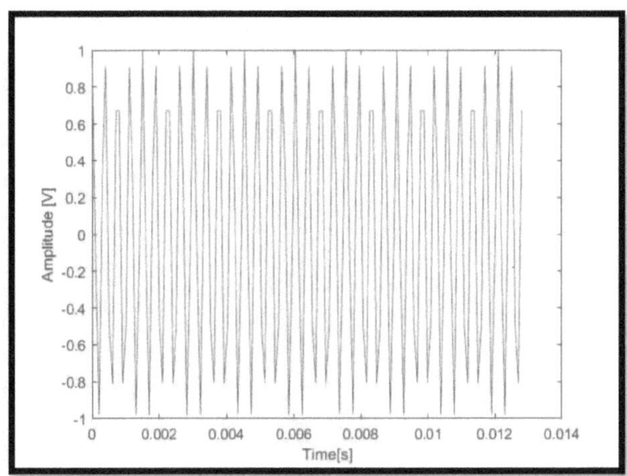

Figure 6-3: Time domain signal of car travelling at 60 km/h in one CPI of 0.0128 s.

The next step in the process was obtaining the 128-point FFT of the signal in Figure 6-3.

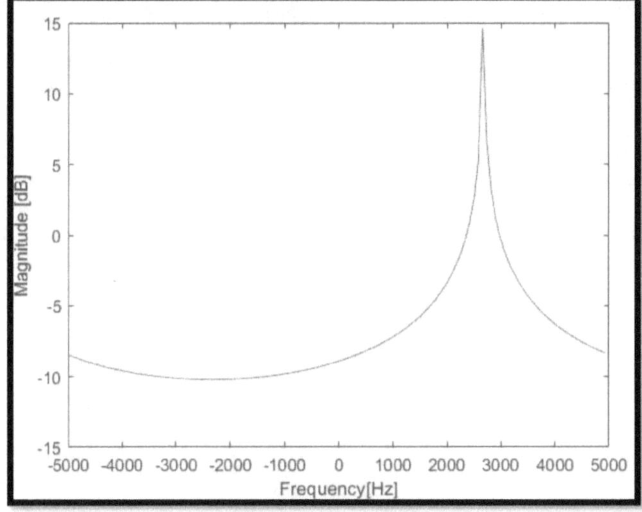

Figure 6-4: FFT of the signal before applying a window

In order to reduce the side lobes/base width, a window function is applied before computing the FFT. This worsened the resolution of the FFT, but the side lobes of the FFT was reduced. This result is shown in Figure 6-5. It can be observed that the frequency resolution of the signal typically broadens by approximately 46% of the unwindowed signal resolution [26]. Which means that after the hamming window was applied, the resolution would be approximately 130.8 Hz.

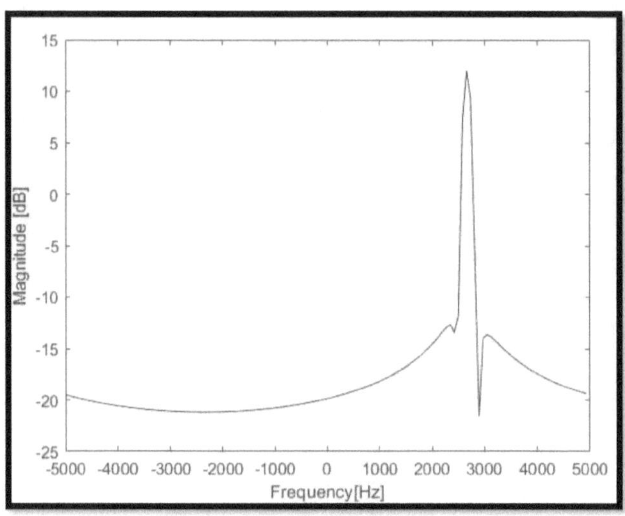

Figure 6-5: FFT of signal after applying a window function

The reduction in SNR was 2 dB, which includes the 1.65 dB straddle loss. The resolution before applying the window function was 89.6 Hz, after applying the window function the actual resolution became 156 Hz resulting in a loss in resolution of about 66.4 Hz.

The next step was applying the CA-CFAR detector. In order to observe realistic results, Gaussian noise was added to the signal. The Gaussian noise is the noise that would typically be found in the receiver of the radar and is caused by thermal noise. A CPI of 0.0512 s was used in the following experiment; this was done to exaggerate the effects of the CA-CFAR. The effects of windowing on reducing the magnitude of the noise and signal peak can be observed in Figure 6-6.

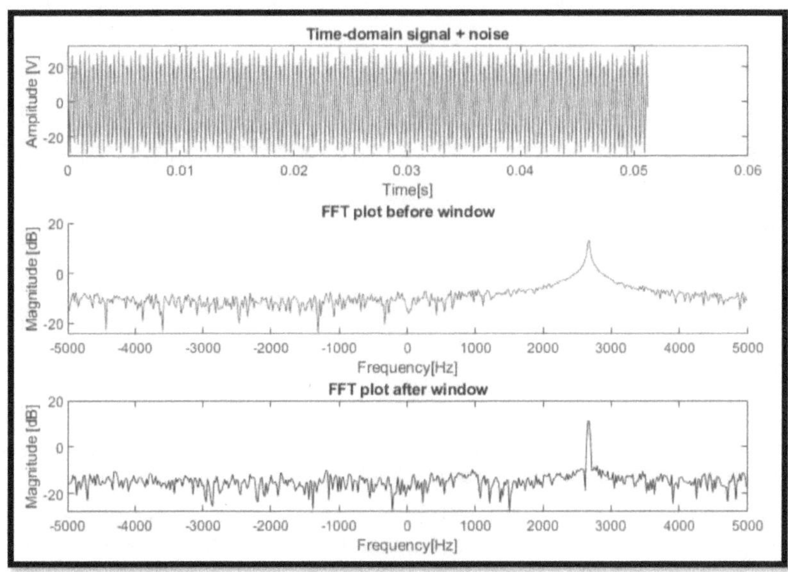

Figure 6-6: Simulated data, used for validating steps in the processing algorithm

Before the CA-CFAR detector was used to obtain detections, the correct operation of the CFAR detector needed to be confirmed; this meant obtaining the effective false alarm rate and the P_{FA} error. In order to obtain the effective P_{FA}, the noise data was collected in the environment the radar was to be placed, clutter was removed in the pre-processing of the data. The 2nd requirement states that there should be only one false detection in a million samples. Thus, a million samples were collected, and the CA-CFAR was used on the collected data to obtain the actual false alarm rate.

In order to obtain statistically relevant information and achieve a P_{FA} error of less than 10%, this experiment was conducted 100 times within 3 hours. A total of 100 million samples were collected, considering that the sample rate was 10 kHz, it took 1.67 minutes to obtain a million samples, and the total experiment took 2.76 hours. Precision was demonstrated when the CA-CFAR was used on the data again, but the P_{FA} parameter was changed at every trial, in order to establish confidence in the system. The P_{FA} values that were used were 1e-6, 1e-5, 1e-4 and 1e-3, the results were summarized in Table 6-1.

This experiment was conducted using a CPI of 9936 seconds which was the total duration of the experiment in seconds.

Table 6-1: Effective false alarm rate vs required of false alarm rate

Coherent Processing interval (s)	Number of Samples	Total number of recordings	Total number	Total number of	Mean false detections/	P_{FA}	P_{FA} error

	Per recording		of Samples	false detections	Effective P_{FA}		
9936	1 Mega Samples	100	100 Mega Samples	108	1.08e-6	1e-6	8%
9936	1 Mega Samples	100	100 Mega Samples	1081	1.081e-5	1e-5	8.1%
9936	1 Mega Samples	100	100 Mega Samples	10810	1.081e-4	1e-4	8.1%
9936	1 Mega Samples	100	100 Mega Samples	108100	1.081e-3	1e-3	8.1%

Table 6-1 shows that there was a total of 108 false detections in 100 Mega samples for P_{FA} of 1e-6; this translates to 939 samples in 24 hours.

In order to obtain data that was representative of the system, different parameters were substituted into the simulation. These parameters included the SNR, the number of reference cells and guard cells. The system required signals with an SNR between 13.19 to 22.5 dB based on Figure 4-2 and Figure 5-13, this informed the decision to use these two values as the minimum and maximum expected signal to noise ratios in the simulation.

The number of reference cells was chosen with the intension to minimize the CFAR loss whilst also avoiding target masking. CFAR loss is defined as the ratio of the SINR required for a CA-CFAR detector to that required for an NP detector, for a given P_D and P_{FA}. Figure 6-7 illustrates the CFAR loss as a function of CA-CFAR window size, N, for three different values of P_{FA} and a 90% P_D.

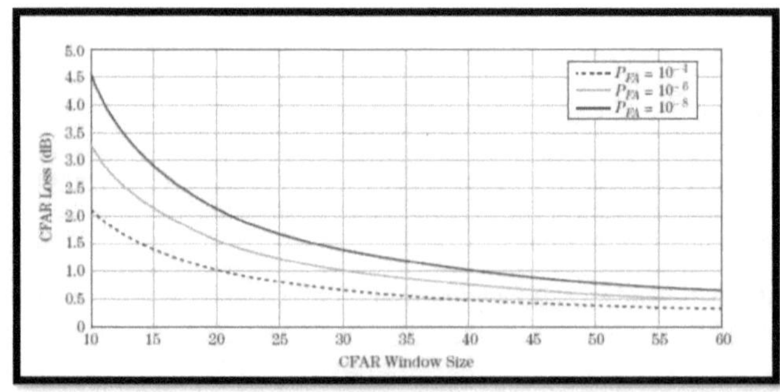

Figure 6-7: CFAR loss as a function of CA-CFAR window size for three different values of P_{FA} and a 90% P_D [26].

his project aims to detect targets at a maximum distance of 40 m; this is in compliance to the road use conditions that states, a vehicle travelling at the maximum speed limit of a particular road must have a minimum of 2 seconds to react to traffic conditions and road signs [3]. Since the maximum speed limit in which the radar sign would be placed in was 60 km/h, the distance in which the radar must detect and display the vehicle speed is at least 33.33 m.

Thus, the SNR must be enough at 40 m in order for the CA-CFAR detector to declare a detection; this means that the system must incur minimal losses. In Figure 4-2, it was shown that the NP detector requires 13.19 dB of SNR to achieve a 90% detection rate given a P_{FA} of 1e-6. Whilst in Figure 5-13 it was shown that a CA-CFAR detector with the same P_{FA} and using N=24, required 22.5dB of SNR to achieve a 90% detection rate. This was a CFAR loss of 1.7 dB. To reduce the CFAR loss to below 1 dB, the minimum number of reference cell required was 30 [26].

Table 6-2 highlights the different parameters that the system was tested against, the number of reference cells were increased by ten reference cells in each test starting with N= 30 and ending at N =50, for SNR of 13 dB, doing so would expose the CA-CFAR threshold behaviour. An optimal number of guard cells must also be chosen to reduce the chances of target self-masking.

Self-masking occurs when the averaged cells of the CA-CFAR threshold contains target returns, which in turn puts a bias in the threshold. Since some target returns lie in more than one resolution cell, without guard cells to offset the bias, the target returns would not be detected, this phenomenon was illustrated in Figure 6-8. The minimum number of guard cells required to avoid self-masking was equal to the target's extent, divided by the resolution size. An equal amount of the determined guard cells must be placed on either side of the CUT [26]. Therefore, an even number of guard cells were tested since the expected target would be a point scatterer with a windowed width. Finding these parameters would ensure that the system was sufficiently characterized.

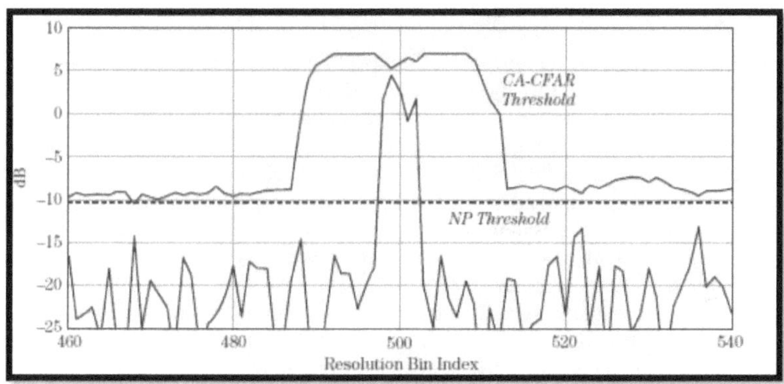

Figure 6-8: CA-CFAR (N=20) with no guard cells exhibits self-masking when an extended target consisting of 5 Rayleigh distributed scatterer is encountered [26].

Table 6-2: Test of different system parameters in simulated data

Parameter	Test 1			Test 2		
CPI	0.0512 s			0.0512 s		
SNR	13 dB			26 dB		
Cases	T1	T2	T3	T1	T2	T3
Leading Cells	15	20	25	30	35	40
Lagging Cells	15	20	25	30	35	40
Guard Cells	0	8	12	16	20	24

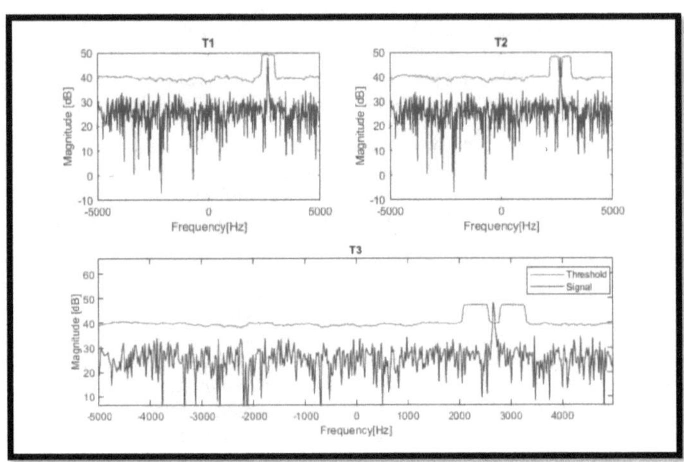

Figure 6-9: Test 1, FFT signal after applying the CA-CFAR when SNR = 13.19 dB

In Figure 6-9 a CA-CFAR detector was applied to the FFT signal with noise when SNR = 13.19 dB, T1 shows a missed detection cause by self-masking since the interference statistic contained the targets returns the threshold around the target contained a bias, which resulted in the target being masked. The target threshold was much higher than the noise signal; this is due to the P_{FA} being low. The threshold smoothed out as more reference cells were used to compute the mean threshold level. T2 and T3 achieved detections, but since T2 had a lower number of guard cells in proportion to the reference cells, only a few samples from the target return registered as true detections. T3 had a total of 12 guard cells and 50 reference cells. Since the CPI for each trial had 512 samples, the signal had a resolution of 19.5 Hz before the window was applied and 28.5 Hz after the window was applied. This result means that after windowing, more of the targets returns will fall in other resolution cells. Hence the target samples become distributed in frequency; thus, it is crucial to have enough guard cells to prevent threshold bias from target returns that have spared to other resolution cells.

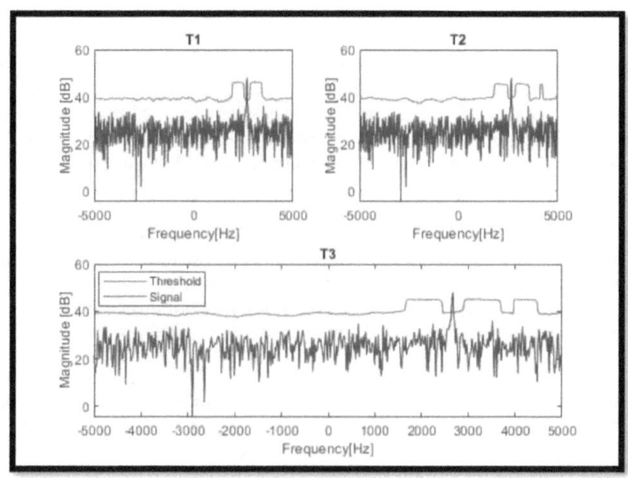

Figure 6-10: Test2, FFT signal after applying the CA-CFAR when SNR = 20 dB

Figure 6-10 illustrates that the more reference cell there are, the smoother the mean threshold level becomes. In T2 and T3, an interesting phenomenon occurred, three threshold peaks in the spectrum are seen. This result was caused by the CA-CFAR algorithm boundary conditions when the index of the CUT was lower than the number of guard cells plus the number of reference cells the threshold is calculated using the references just before CUT.

Since the reference cells before the CUT contained the target returns, the threshold contained the bias, resulting in an extra raised region in the threshold.

This would be a problem if another target was travelling at the velocity corresponding to that region and is referred to as target masking. The optimal number of the reference cell and guard cells was then concluded to be 50 and 12, respectively, these parameters are found in Test 1 T3. These parameters correspond to a CFAR loss of less than 0.5 dB and no self-masking.

6.2.2   Analogue to Digital Converter

The ADC that was chosen for the data-collection system is the Picoscope 2206B oscilloscope, this system has the following specifications

Table 6-3: Picoscope 2206B Specifications

Parameter	Symbol	Value	units
Resolution	b	8	bits
Sampling Rate	f_s	100	kHz
Bandwidth	B	50000	kHz
Waveform generator		Y	Y/N
Power Consumption	P	2.5	W

The Picoscope must be able to sample small amplitude returns. Therefore, the lowest amplitude signal the ADC can represent without adding too much quantization noise must be simulated. When the simulated signal is obtained, the Picoscope must then sample a real signal with the same characteristics (Doppler frequency and amplitude) as the simulated signal. The two results must be compared, and if the spectrums are similar, then the Picoscope shall be declared fully functional.

In this experiment, the signal the digitized must have the same frequency associated with the highest anticipated velocity, which is 60 km/h. This is to ensure that the system can digitize signals with frequencies that correspond to those stated in the requirements. The experiment was first to be simulated using MATLAB; the results of the simulation were then compared with an experiment using real signals. These real signals were generated using the arbitrary waveform generator of the Picoscope.

A sinusoid with a frequency of 2666.67 Hz was simulated using MATLAB, calculated using Equation 3-1, $x(t) = \{ \cos(2\pi f_0 t)$, and amplitude of 20 mV and a CPI of 0.0512 s.

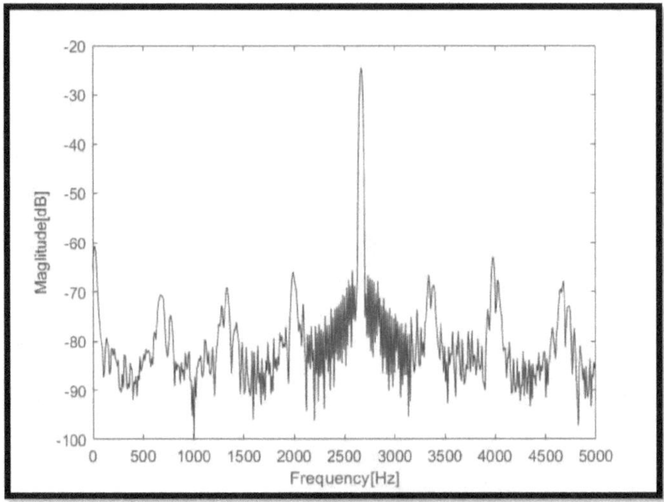

Figure 6-11: Magnitude spectra of 20mV quantized signal.

Figure 6-11 shows the anticipated magnitude spectra of a 20 mV quantized signal that has a frequency of 2666.67 Hz. The magnitude spectrum has pronounced frequency harmonics that are caused by the quantization noise as described by the simulations done in Appendix A.2, the more resolution the ADC has, the less pronounced the harmonics which is in line with the theory presented in Section 3.6.

The simulated experiments shown in Figure 6-12 and Figure 6-13 were then made using the Picoscope's arbitrary waveform generator and the FFF functionality.

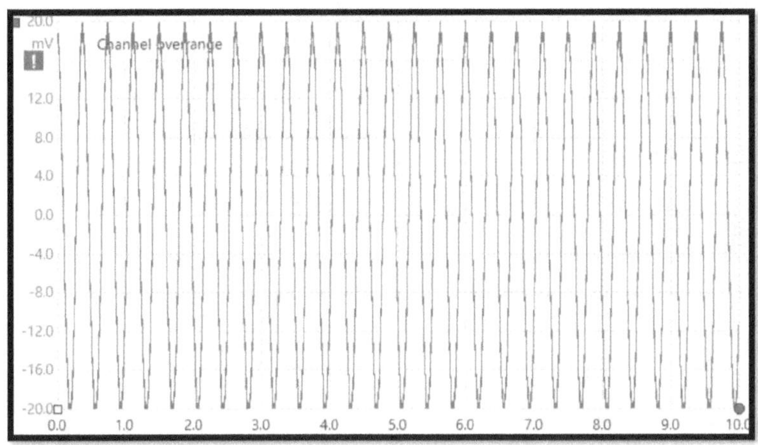

Figure 6-12: Quantized 20mV signal.

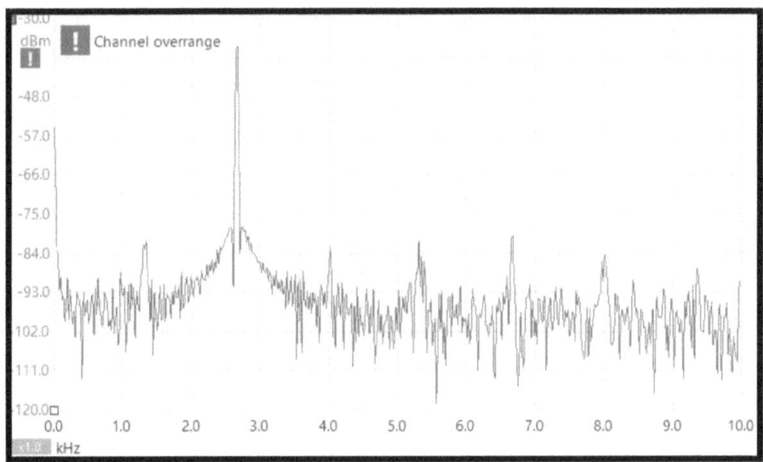

Figure 6-13: FFT of 20 mV quantized signal.

Figure 6-13 shows the expected magnitude spectra for the input sinusoid with an amplitude of 20 mV. The frequency harmonics presented in Figure 6-13 are comparable to the frequency harmonics present in Figure 6-11. It can be observed that the ADC receiver collected samples that are greater than 20 mV; this cannot be a possible explanation to the harmonics in Figure 6-13. These frequency harmonics might be due to the power supply unit as frequency harmonics arise when the signal level of the receiving channel is not connected to same ground level as the AWG, this hypothesis was confirmed in Section 6.2.5.

The amplitude of the signal in Figure 6-13 was in dBm and a 600Ω resister to obtain the scaling whilst the amplitude of the signal in Figure 6-11 was in dB. The main lobe amplitude peaks relative to their frequency harmonic amplitude peaks are comparable in magnitude, as the main lobes in Figure 6-11 and Figure 6-13 are about 40 dB greater than the amplitudes of the harmonic distortion amplitudes.

6.2.3 Radar Module

In this section, the radar module will be evaluated to confirm correct operation. The application document provided by the manufacturer shows the typical experimental set up to determine if the module is functioning correctly.

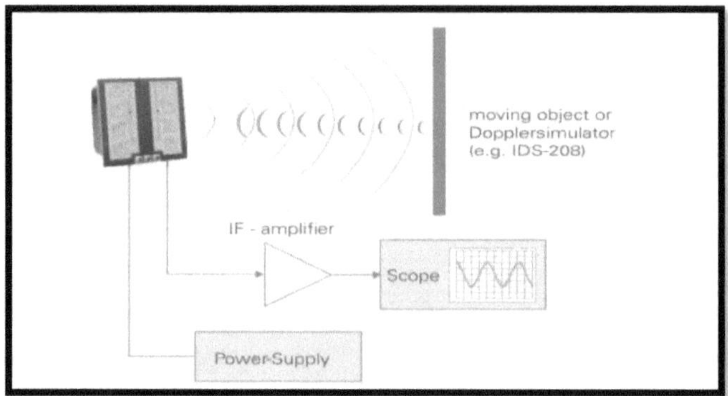

Figure 6-14: Experimental set-up for radar module test [29].

Figure 6-14 shows the experimental set-up to determine if the radar module is functioning correctly. The equipment used was an oscilloscope and a power supply. The Picoscope functions as a portable oscilloscope and the laptop USB port provides 5 V DC and up to 500 mA. The Innosent IPS-154 specifications are outlined in Section 5.2 Table 5-2. The first experiment to be carried out is to determine the noise floor of the system. This experiment was carried out under laboratory conditions; the result of this experiment is Figure 6-15. In Figure 6-15 it is observed that there are no sources of interference , the noise appears to be uniform.

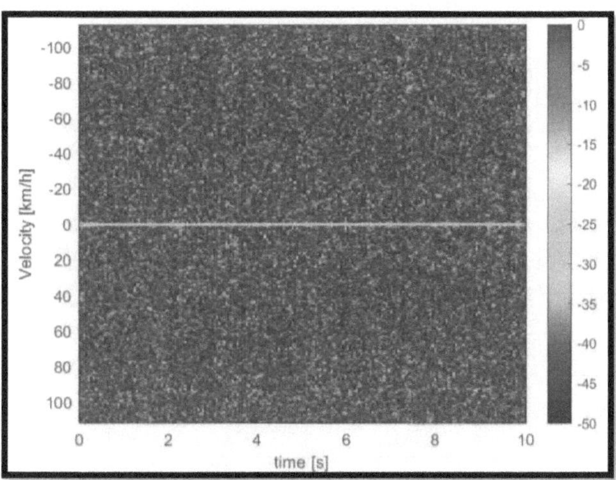

Figure 6-15: A spectrogram of the noise measured in the laboratory.

The second experiment consists of a hand being swung back and forth 60 cm away from the IPS-154 sensor. The hand will be moving away and towards the sensor in a short space of time, resulting in the positive and negative Doppler return over time seen, Figure 6-14. Figure 6-16 is the results of the experiment mentioned above. The sinusoidal frequency content in Figure 6-16 is the same as the frequency content in the illustration shown in Figure 6-14.

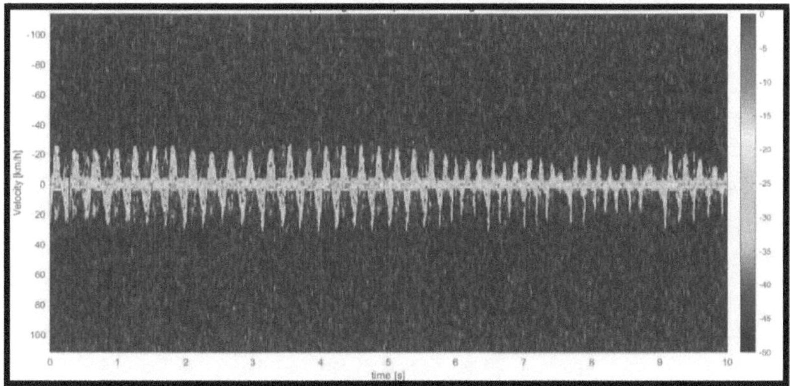

Figure 6-16: A spectrogram of a hand waving rapidly.

It can thus be concluded based on the results shown in Figure 6-15 and Figure 6-16, the radar module is operating as expected as detailed by the IPS-154 datasheet [29].

6.2.4 Display and Power System

The experiments conducted in this chapter will utilize a Samsung Series 7 Chronos Laptop as both a power source, DSP and display. The specifications of this laptop are detailed in Table 6-4.

Table 6-4: Specifications for Samsun Series 7 Chronos

Parameter	Specification
Central Processing Unit (CPU)	Intel Core i7-3635QM CPU
CPU Speed	Typ. 2.4GHz Max. 3.4GHz
Graphics	NVIDIA GeForce GT 640M 1GB
Graphics card processing speed	20.00 GFLOPS
Random Access Memory (RAM)	8GB DDR3 memory
Hard Drive Disk (HDD)	1TB SATA II (5400 rpm)
Display type	High Definition (HD) Liquid Crystal display (LCD)
Display Size	15.6 inch
Display Brightness	300 nits
Power input	19 V DC 3.16A ~60W
Power output (per USB port)	2.5W

The graphics card speed is rated at 20 GFLOPS, meaning it will be able to handle the FFT multiplications required for this application when operating MATLAB using the graphics processor. It will take this processor 89.6 ns to finish 1792 multiplications required for a 128-point FFT [59]. The laptop can provide a constant 2.5 W of power per USB port. The ADC and radar module are connected on separate UBS ports.

6.2.5 Integration Issues

In integrating these sub-systems, a few challenges arose. The challenges faced in the testing and integrating of these sub-systems were with regards to signal interference from unknown sources. These issues are detailed in Table 6-5.

Table 6-5: Integration issues and possible solutions

Sub-system	Issue	Possible solutions
IPS-154	Unknown interference	Perform spectral analysis to obtain the frequency components of the interference and develop a digital filter to attenuate the unwanted signals.

The main challenge encountered in this project was due to an unknown source of interference when measuring the noise floor in the laboratory. Consider Figure 6-17, where interference signals can be observed at ± 4 kHz or ± 90 km/h and ± 2 kHz or ± 45 km/h.

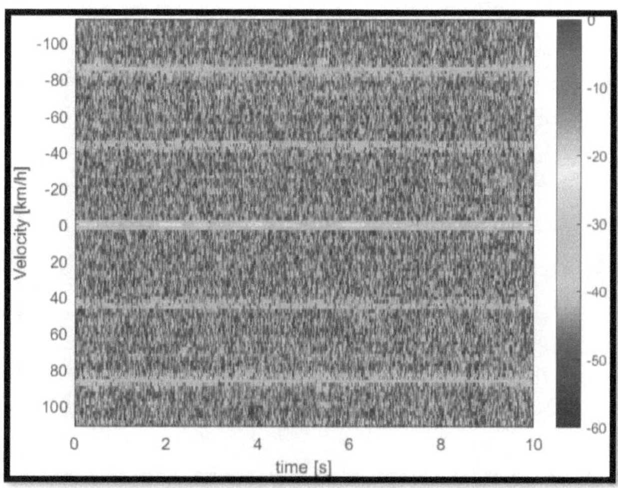

Figure 6-17: A spectrogram of an unknown interference signal.

The interference bands persist for the duration of the recording, and upon further inspection, the signals seem to be mirror images of one another. This result meant that both the I and Q channels were receiving the same signal interference; this was found to be the result of inconsistent system grounding. The system operates using the same power source, but it has different voltage levels. Thus, a common ground is required to block frequency harmonics that are associated with the oscillators found in either the radar module or Picoscope. This recommendation was applied when continuing further with the subsequent experiments.

A second abnormality was discovered during initial radar transceiver testing. The rapid hand waving experiment was first performed with the hand at 30 cm away from the device. Consider Figure 6-18, instances of interference at all frequencies can be observed for a limited time; this means that external electronics were not the cause of the interference and has to do with the samples taken at those instances.

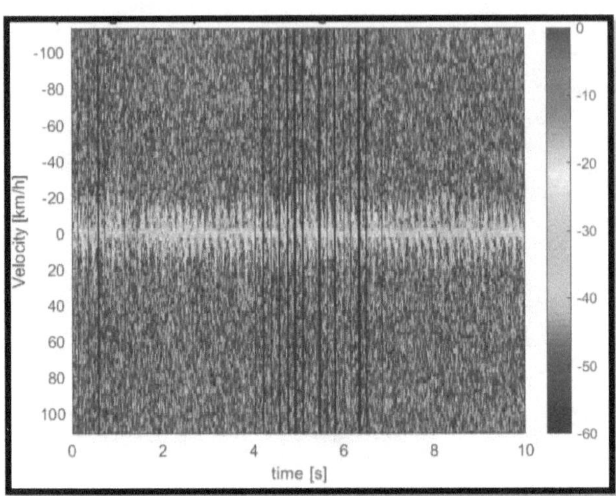

Figure 6-18: A spectrogram of rapid hand waving with unknown interference.

Since the rapid hand waving experiment was conducted very close distance to the radar module, any vibrations caused by the shaking of the sensor could have resulted in no samples being collected for those milliseconds. The lines of discontinuation in Figure 6-18 may be due to Picoscope assigning a default value of "-inf" to those instances [45]. Figure 6-18 was created using the imagesc function from MATLAB when this function is a given sample with value "-inf" indicating negative infinity; it attributes it the lowest pixel value of the colour map; this results in the lines of discontinuation seen on the plot [33]. This problem may be resolved by ensuring that the connections between the ADC and radar module are robust and do not disconnect even when vibrations occur.

The last challenge was enabling near-real-time operation. The radar signal processing code used the Python language, which supports Picoscope drivers but no streaming function to enable continuous processing of the data. The first thing to note is the processing parameters. The sampling frequency of the radar as stated in the system design is 10 kHz, and because there are 128 FFT points in one Doppler profiles to enable a velocity estimate of $\pm$ 2km/h, thus one coherent processing interval (CPI) was 0.0128 seconds long. The following code snippet shows how this challenge was solved.

```python
def PS2000a():

    print "Attempting to open..."

    ps = ps2000a.PS2000a()

    t = time.time()

    plt.ion()

    fig=plt.figure()

    while True:

        obs_duration =0.0128 #CPI

        sampling_interval =1.28e-6 #obs_duration
/ 10000

    ps.close()
if __name__ == "__main__":

    PS2000a()
```

The code shows that the loop was set up using a while statement. The condition to enable continuously, uninterrupted processing was the addition of the True Boolean condition. This condition ensured that the data is captured per CPI, stated as the obs_duration in the code; this means that 78 Doppler profiles are processed every second since $1/0.0128 = 78$. This result creates a near real-time processing algorithm and a precise velocity estimate. There was no overlap in the processing of the Doppler profiles.

6.3 Summary

In this chapter, the integration and testing of the data-collection system and signal processing algorithm was detailed, and the specifications of each sub-system were outlined. The testing of each sub-system revealed some issues that were a result of inconsistent grounding and missed samples. The issue of inconsistent grounding was rectified by connecting a single ground wire to the negative terminals all the 12 V DC powered components. The next issue was rectified by ensuring that the connections between the ADC and radar module are robust and do not disconnect even when vibrations occur. The last issue experienced was that of continuous capturing and processing of the radar data. This was rectified by using a while statement with a True Boolean condition. The signal processing algorithm produced a consistent result that agreed with the theory studied in Section 5.4. The next chapter is the acceptance testing and results of the experiments.

Chapter 7

Acceptance Testing and Results

In the previous chapter, it was determined that each sub-system was working as intended through analysis of simulated data and data measured under lab conditions. In this chapter, the experiments designed in Section 4.4 Table 4-2 were carried out using the data-collection system. The results of the experiments were outlined as well as analysed. The quality of these results would also determine the suitability of the proposed radar solution. The first experiment in this chapter was done to obtain the mean noise intensity, which was used in characterizing the achievable SNR of the system. The second experiment involves obtaining the experimental SNR of the system at specific distances using the smallest vehicle found on the business campus. The smallest vehicle was used because the minimum detectable SNR of the system was required. The third experiment was done to verify the system accuracy and the minimum detectable distance of the system; this was achieved by driving the vehicle with the electronic cruise control programmed to drive the vehicle at a constant speed of 20 km/h and 40 km/h respectively. The last experiment quantified the maximum detectable speed that the radar can discern; this was done by accelerating the vehicle from a considerable distance so that when the vehicle reaches 40 m, it would be travelling at 60 km/h.

7.1 Experiment A1: Noise Intensity

This experiment involves placing the radar in a typical environment of operation to determine the average noise levels that the system would be exposed to whilst operating on a day-to-day basis. The scene consisted of foliage, an asphalt road and a large building. The returns from these objects are anticipated to be found in the zero Doppler region of the spectrogram, referred to as the clutter region since the objects are stationary.

Figure 7-1: Experimental set-up for Experiment A1.

Figure 7-1 shows the radar placed on the side of the road facing oncoming traffic. Thermal noise would be the primary cause of noise in the system since the radar would be placed in direct sunlight. This noise was assumed to be homogenous or normally distributed. In order to prove this assumption, the means of the noise samples must be calculated and by the Central Limit Theorem (CLT), the distribution of the means must approximate the shape of a normal distribution.

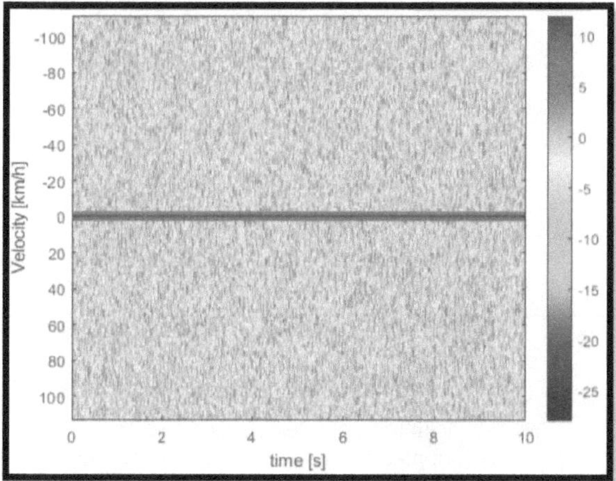

Figure 7-2: Spectrogram of noise in Experiment A1.

From Figure 7-2, specific regions were analysed in Doppler profiles from 20km up to 60km/h in both directions. The zones from 2km/h up to -2km/h were excluded from analysis as the clutter resides in this region. There are two noise regions that were analysed. The positive velocity noise region which shall be referred to as Noise+ and the negative velocity noise region which shall be referred to as Noise- and the of the two sides are then concatenated to produce the total noise of the noise sources which is the noise floor. Refer to the signal processing code in Appendix A.3, for the implementation details.

The following code was used to compute the Mean Noise intensity of receiver noise.

```matlab
% % %% Mean Noise Intensity calculation
Start = 1; % Matlab starts calculating from the first index
V_pos_start = 20; % Positive input speed in km/hr
V_pos_finish=60;

V_neg_start = -60; % Negative input speed in km/hr
V_neg_finish=-20;

Number_of_top_half_samples = 64;
Number_of_bottom_half_samples = 64;

Highest_neg_velocity = -112;
Highest_pos_velocity = 112;

Neg_Noise_sample_start= Number_of_top_half_samples -
V_neg_start*(Number_of_top_half_samples/ Highest_neg_velocity)

Neg_Noise_sample_neg_finish = Number_of_top_half_samples -
V_neg_finish*(Number_of_top_half_samples/ Highest_neg_velocity)

Pos_Noise_sample_start= Number_of_top_half_samples + V_pos_start
*(Number_of_bottom_half_samples / Highest_pos_velocity)
Pos_Noise_sample_finish= Number_of_top_half_samples + V_pos_finish
*(Number_of_bottom_half_samples / Highest_pos_velocity)

Noise_negetive = CW_DATA_win (Neg_Noise_sample_start: Neg_Noise_sample_neg_finish
,Start: nDopProfiles);
Noise_positive= CW_DATA_win(Pos_Noise_sample_start:Pos_Noise_sample_finish,Start:
nDopProfiles);

Noise_full =cat(1,Noise_negetive,Noise_positive);
mean_CPI =zeros(1,length(nDopProfiles ));

for i =Start: nDopProfiles

    mean_CPI(i)= mean(abs(Noise_full(:,i)));
end
histogram(mean_CPI,30,'Normalization','probability');
xlabel('Magnitude')
ylabel('Probability ')

Total_Noise_intensity = sum (mean_CPI)/nDopProfiles;
Total_Noise_intensity_dB = 10*log10 (Total_Noise_intensity);
```

The code above shows how the mean noise intensity was calculated. The parameters that were varied were the samples per Doppler profile, effectively increasing the CPI. Since system calibration was not possible, the total noise power could not be calculated; instead, the mean noise intensity was calculated and converted into dB for simplicity. The mean noise intensity estimate improved the more samples were used; this observation was consistent with theory. The results were summarized in Table 7-1.

Table 7-1: CPI vs Mean Noise intensity in dB

Samples per Doppler profile	Coherent Processing interval (s)	Mean Noise intensity (dB)
512	0.0512	1.8
256	0.0256	1.10
128	0.0128	0.9

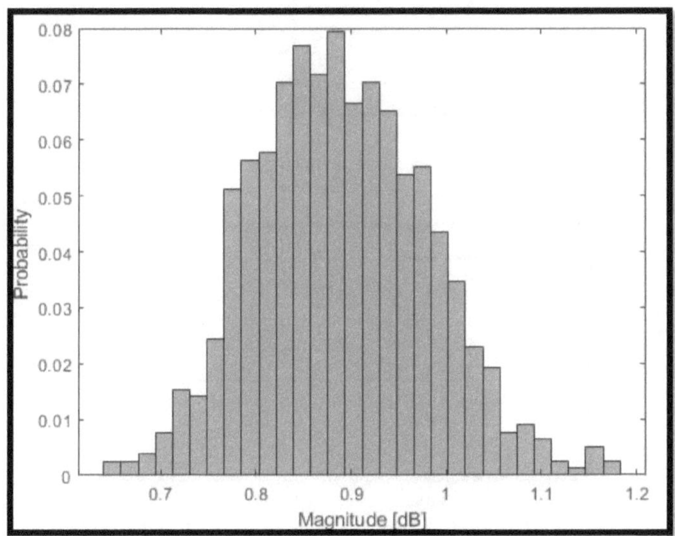

Figure 7-3: Histogram of the means of the noise samples for CPI =0.0128 s

Figure 7-3 shows the histogram of the means of the noise samples, which represent the probability distribution function (PDF); the shape of this graph was similar to the Normal distribution. Therefore, by the central limit theorem, the noise can be approximated as being Gaussian and homogenous.

7.2 Experiment A2: Experimental SNR

The previous experiment determined the mean noise intensity of the scene, and it was established that the assumption that the noise is Gaussian and homogenous was correct. The values of noise intensity were used to determine the experimental SNR as a function of range.

Figure 7-4: The experimental setup to determine the received signal power of a typical vehicle

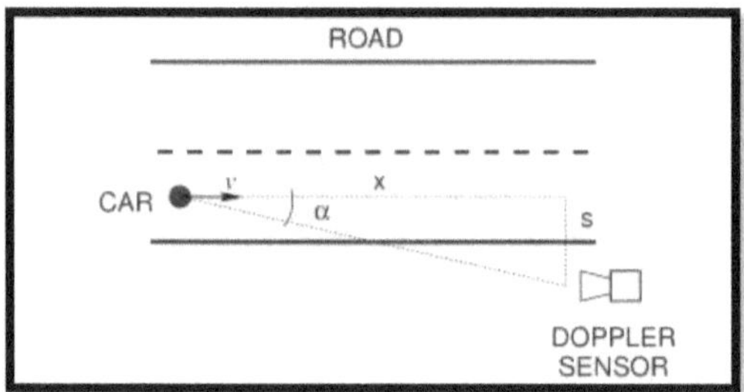

Figure 7-5: Top view illustration of the experimental setup [16].

The experimental setup consists of a small hatchback vehicle approaching the radar system from 40 m at a constant speed of 20 km/h. When the car was at 12 m from the radar, it started to decelerate to a complete stop. The SNR of the vehicle return was expected to increase as the vehicle approaches the radar system since SNR is inversely proportional to the fourth power of the range, i.e. SNR $\propto \frac{1}{R^4}$, when converting this relationship into the dB scale, the relation becomes $SNR_{dB} \propto -4\log_{10} R$.

The parameters relevant to this experiment were:
- Initial Range = 40 m
- Number of Samples in a Doppler profile = 128, 256,512
- Initial velocity = 20 km/h
- Mean Noise Intensity = 0.9, 1.1, 1.8 dB

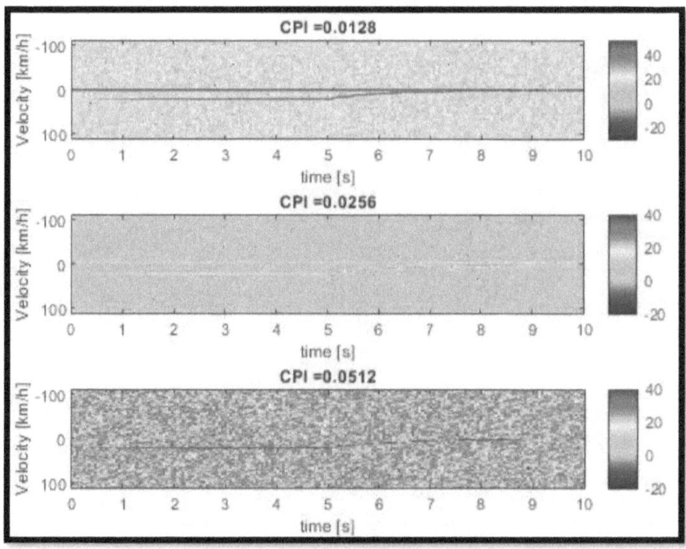

Figure 7-6: Vehicle approaching towards the radar at 20km/h then decelerating to 0km/h after 8 seconds.

Figure 7-6 shows spectrograms of a vehicle travelling towards the radar system. The intensity of the spectrogram line associated with the vehicle increases as the vehicle gets closer to the radar system, as expected. In order to quantify this observation, the SNR vs range plot was made; by finding the maximum intensity in a Doppler profile then calculating the ratio between it and the of the noise intensity, for every Doppler profile in the spectrogram.

Figure 7-7 shows the relationship between the SNR as the vehicle approaches the radar, the expected relationship was observed, i.e. as the vehicle gets closer to the car ($R \to 0$).

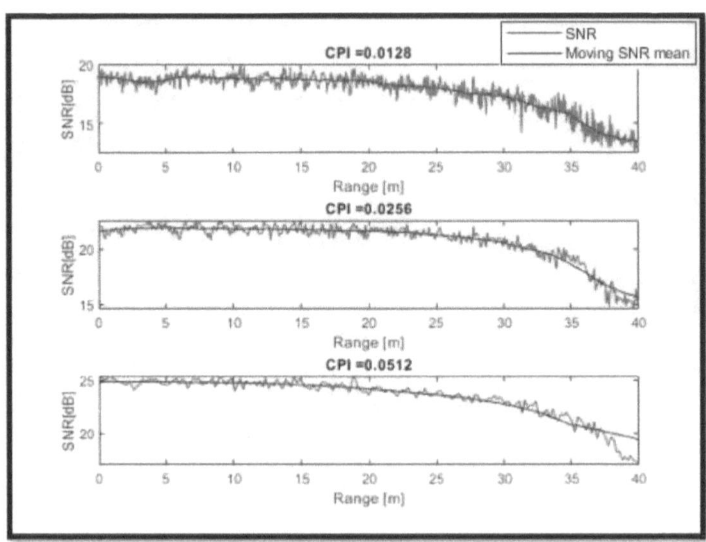

Figure 7-7: SNR as a function of range

The relationship between the SNR and the range was explored in this section, the assumption that range is a product of velocity and time was made in order to convert from the time axis to the range axis. In order to obtain the correct scaling for the data observed, an initial distance of 40 m was used.

This assumption led to Figure 7-7, which shows the SNR at each range as the car approached the radar for different CPIs. The red line shows the actual SNR readings whilst the blue line is the moving mean of the SNR graph when using a sliding window consisting of 50 samples; there was no overlap with the successive blocks of data. This moving average was used to illustrate the general trend of the data in order to verify the relation $SNR_{dB} \propto -4 \log_{10} R$. It was observed that the SNR increased when the CPI was increased, this relationship was expressed in Equation 3-3, $SNR = \frac{N\sigma P_t G_t G_r \lambda^2}{(4\pi)^3 R^4 P_n}$, wherein the case of CW radar, N is the number of samples collected in a Doppler profile. The longer the CPI, the more samples are collected and added coherently since the phase information of the signal is preserved. The mean SNR when the CPI is chosen to be 0.0128 s, 0.0256 s and 0.0512 s is 18, 23 and 25 dB respectively.

In order to contextualise these results, the expected dynamic range for the system was calculated; given the Picoscope's specifications. In Section 5.3 Table 5-3, the manufacturers specified a spurious-free dynamic range (SFDR) of less than 44 dB @10 kHz when using the channel limits of ±20 mV. In Section 5.2.3, the expected SNR for the system using the module IPS-154 was calculated to be 50.57 dB at 40 m; this result excludes the system losses L_s [26].

$$L_s = L_t L_a L_r L_{sp}$$ 7-1

Where

L_t: Transmit loss

L_a: Atmospheric loss

L_r: Receiver loss

L_{sp}: Signal processing loss

The losses associated with the radar module such as the transmit loss and receiver loss are found in the IPS-154 data sheet [37] ; the transmit loss consists of the power lost due to spurious emissions as well as power taped by the coupler. The receiver loss is due to the mixer conversion process where the signal is mixed down to IF. The signal processing loss consists of the CFAR loss and windowing loss; these values were calculated and found to be 0.5 dB and 1.68 dB, respectively. The total system losses were found to be 16.3 dB; thus, the DR of the system is 34.27 dB.

7.3 Experiment B1: System Accuracy and Maximum Detectible velocity

In order to determine the velocity measurement accuracy of the system, a reference system was proposed in which the velocity measurements were compared. The reliability of the reference system needs to be confirmed before the experiments can be carried out.

To obtain the accuracy of a quantity, a control or a true measurement must be established before the experimental measurements can be made. This is establishing ground truth for accurate measurement accuracy analysis. There are several ways to establish ground truth, one way to establish ground truth of a Doppler measurement is by using a global positioning system (GPS). Advanced GPS systems such as the Advance Navigation Spatial Dual system uses a global navigation satellite system (GNSS), which consists of several satellites in space that broadcast navigation signals. This system provides an accurate position of up to 2.5 m and velocity accuracies of up to 0.03 m/s or 0.108 km/h [60].

Unfortunately, this system was not available at the time the experiments were being carried out. This project made use of an electric vehicle that has a very accurate speedometer. The BMW i3 hybrid electric car was used for this project.

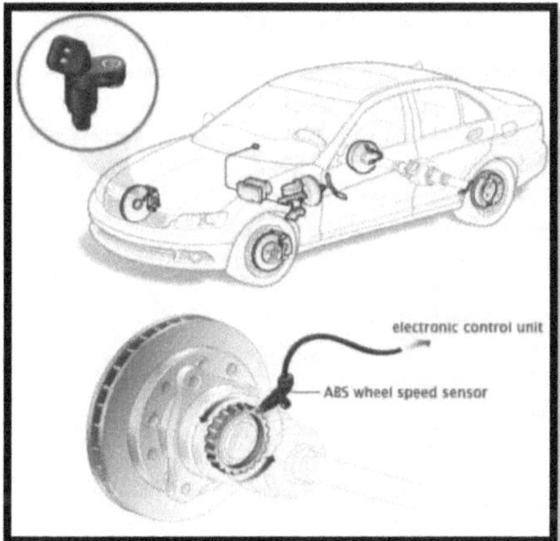

Figure 7-8: Electric speed sensor. The illustration depicts where the ABS wheel speed sensor resides in all ABS vehicles [61].

The BMW i3 vehicles use the Anti-lock braking system (ABS) wheel speed sensors to measure the speed of the vehicle. Active sensors in the BMW i3 are more accurate and are able to detect speeds of less than 0.09656064 km/h, which was more accurate Advance Navigation Spatial Dual system[61]. This level of accuracy is required to establish a velocity measurement truth value for experiments in the radar measurements. The added advantage of using the BMW i3 electric car was that it had built-in controls to allow for a constant continuous speed. These features enabled the Experiment B1 to be carried out. This experiment requires the car to travel at a constant speed towards the radar. The radar must then produce detections that match the speed of the car accurately and precisely.

The experimental set up consists of the BMW i3 40 m away from the radar. The car travels towards the radar at a constant speed of 20 km/h. The radar stopped the measurement once the vehicle was 1 m away from the radar, as shown in Figure 7-9.

Figure 7-9: Experimental set up for Experiment B1.

The resulting spectrogram from one measurement was shown in Figure 7-10, when the P_{FA} was set to 10e-6.

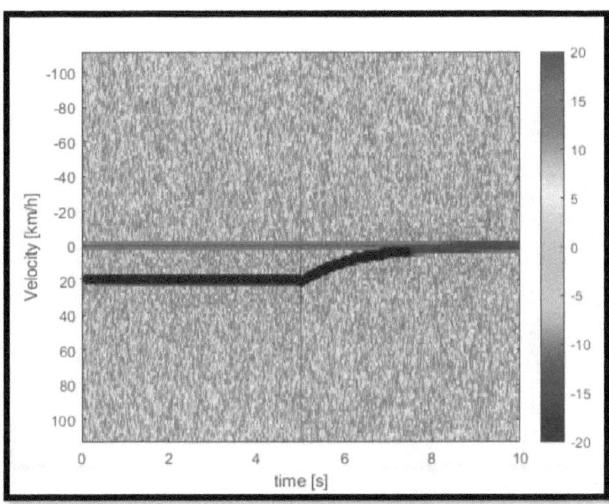

Figure 7-10: A spectrogram of an electric car travelling at constant speed of 20km/h for 5 seconds and then decelerating for 3 seconds to a complete stop, using CPI =0.0128s with N = 50, G =12.

From Figure 7-10, it can be observed that the constant velocity detections start from 0 seconds up until the 5th second of the measurement. This result means that for 5 seconds, the vehicle travelled at a constant 20 km/h, which is a total distance of 27.78 m. The spectrogram also shows the car decelerated to a halt after 3 seconds. The distance the car travelled before it came to a complete halt was then found to be 11.22 m since the car stopped 1m from the radar; this means that the maximum detection distance for a CPI of 0.0128 s was 40 m. In these regions and the entire recording, no false detections were observed, and all visible vehicle returns were correctly identified as detections by the CA-CAFR; this amounts to a 100% detection rate. A P_{FA} of 10e-6 significantly reduces the likelihood of a false detection appearing within the 100k samples. The SNR of the first detection was found to be 13.2 dB as seen in Figure 7-7, which means there was sufficient SNR for a PD of 90% for an NP detector, and since the car was approaching the radar, the SNR increased which also improved the probability of true detections being made.

The region between -2 km/h and 2 km/h was not analysed; this allowed the detection algorithm to reject any detections that corresponded to velocities lower than that limit. The velocity estimate that corresponds to the first detection was shown in Figure 7-11. The signal amplitude was found to be 33.25 dB, and the average noise levels were at 20 dB. The reference cells used reduced the CFAR loss to less than 0.5 dB; this resulted in the SNR of the first detection being 13.25 dB, which is consistent with Theoretical levels predicted in Figure 4-2 by the NP detector.

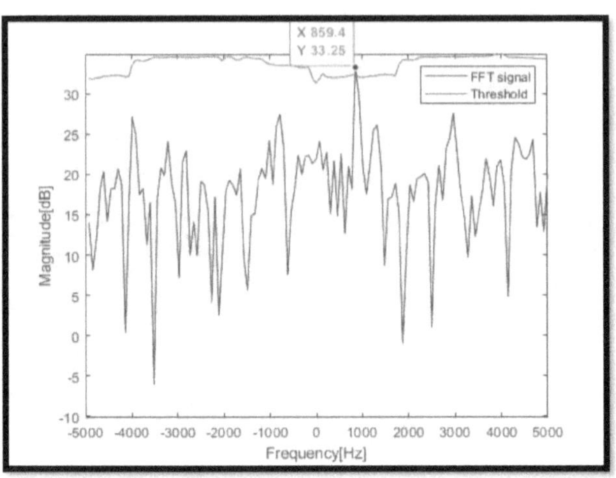

Figure 7-11: 1st detection by the CA-CFAR with N = 50, G =12 for a CPI of 0.0125s

Figure 7-11 illustrates that the radar could detect a speed of 19.33 km/h. This estimate was, however, limited by the Doppler resolution of the system which was dictated by the FFT-length. In order to obtain the actual measurement estimate, data interpolation was used to obtain the true measurement velocity estimate of the vehicle. This process required the detections to be saved in detection matrix; this matrix contained detections from the 1st Doppler profile up until to the 390th Doppler profile when using a CPI of 0.0128 s, this spans the time the vehicle was travelling at a constant speed.

The detection matrix elements in each Doppler profile dimension were added in order to reduce the matrix into a detection vector. Figure 7-12 was the result of using interpolation on the data by the 'spline' method. The velocity estimate was then calculated by finding the peak value of the graph. This process resulted in a velocity estimate of 19.96 km/h.

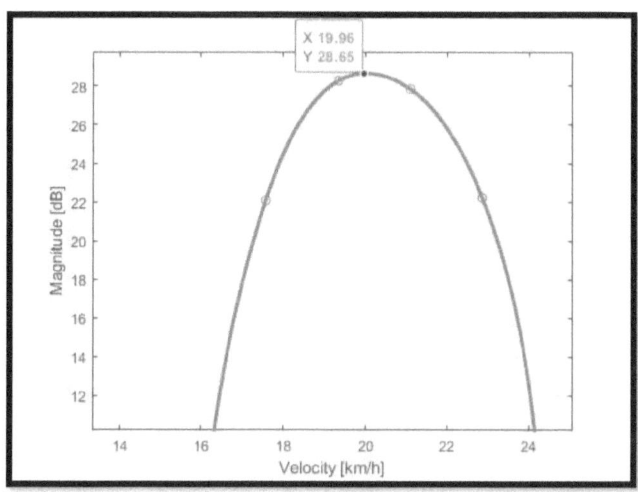

Figure 7-12: Velocity estimate using interpolation for a CPI =0.0128 s.

The process was repeated on the same data but using a CPI of 0.0256 s; this was done to see if the velocity estimate and the detection distance would improve since it was postulated that the velocity estimate accuracy would increase with a high CPI. Since the detection distance was 40 m, there would likely be no further improvement, but there must be no degradation in the system performance as well.

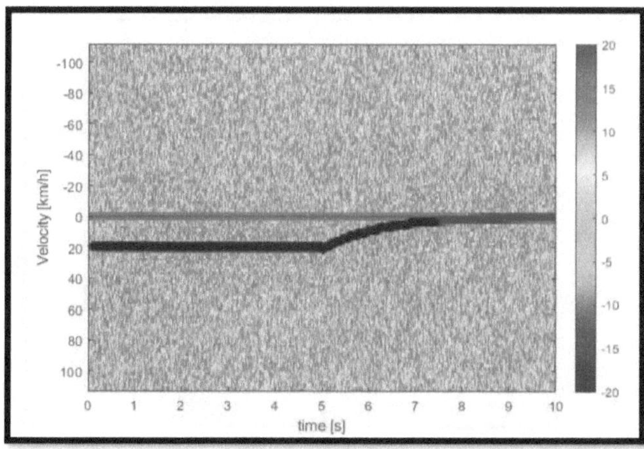

Figure 7-13: A spectrogram of an electric car travelling at a constant speed of 20 km/h for 5 seconds and then decelerating for 3 seconds to a complete stop, using CPI =0.0256 s with N = 50, G =12.

Figure 7-13 shows that the car was detected at 0 s, this means for a car travelling at 20 km/h with CPI of 0.0256 s the maximum detectable target distance was 40 m, which means the detection distance did not change from using a CPI of 0.0128 s, the CA-CFAR still achieved 100% detection rate and no false alarms. The SNR of the first detection was found to be 14 dB, which was calculated using the signal magnitude as well as the mean noise levels from Figure 7-14.

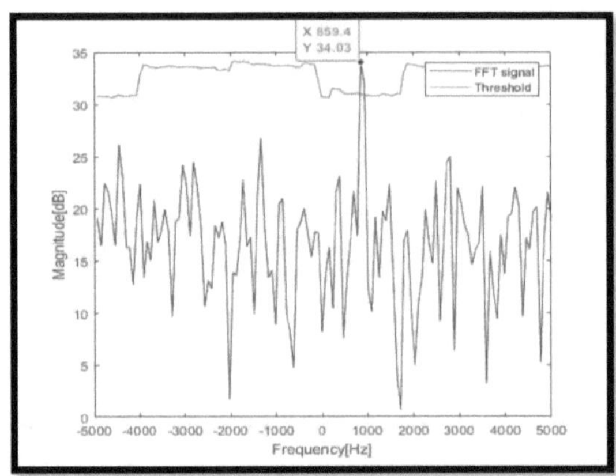

Figure 7-14: 1st detection by the CA-CFAR with N = 50, G =12 for CPI of 0.0256s

It can be seen in Figure 7-14 hat the signal magnitude has increased to 34.03 dB and the mean noise levels remained at 20 dB. The velocity estimate, which is limited by the Doppler resolution, was found to be 19.34 km/h. The same method of interpolation using the 'spline' method was repeated to find a more accurate velocity estimate, and the results are shown in Figure 7-15.

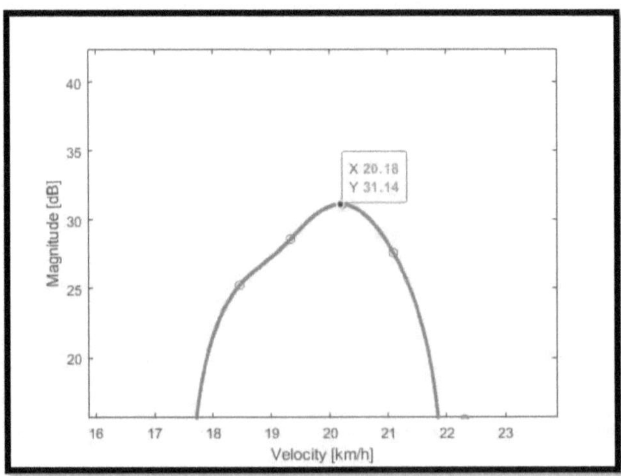

Figure 7-15: Velocity estimate using interpolation for CPI =0.0256s.

Figure 7-15 shows the velocity estimate using interpolation for a CPI of 0.0256s, was found to be 20.18 km/h. The velocity estimate graph that was found using interpolation had more data points to extrapolate from; this stems from the fact that the resolution of the system was finer.

The process was again repeated using the same data but changing the CPI parameter to 0.0512 s, the detection distance remained 40 m, and the CA-CFAR maintained a 100% detection rate whilst not producing any false detections. The artefact shown at 5s was a result of the vehicle abruptly decelerating, which resulted in the car coming to a complete stop after 3 s.

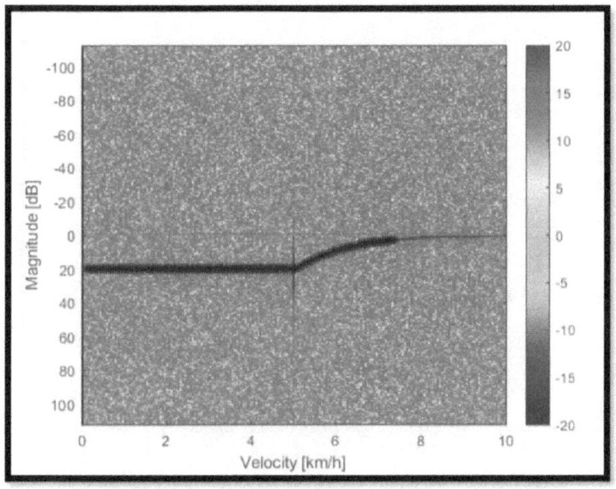

Figure 7-16: A spectrogram of an electric car travelling at a constant speed of 20km/h for 5 seconds and then decelerating for 3 seconds to a complete stop, using CPI =0.0512 s with N = 50, G =12.

A velocity estimate of 19.78 km/h was found with a corresponding SNR of 22.09dB. This was shown in Figure 7-17.

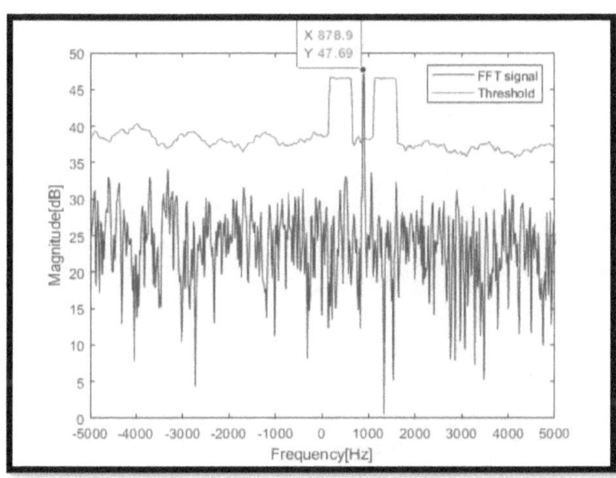

Figure 7-17: 1st detection by the CA-CFAR with N = 50, G =12 for CPI of 0.0512 s

After interpolating and finding the peak value of the graph, a velocity estimate of 20 km/h was found, as shown in Figure 7-18.

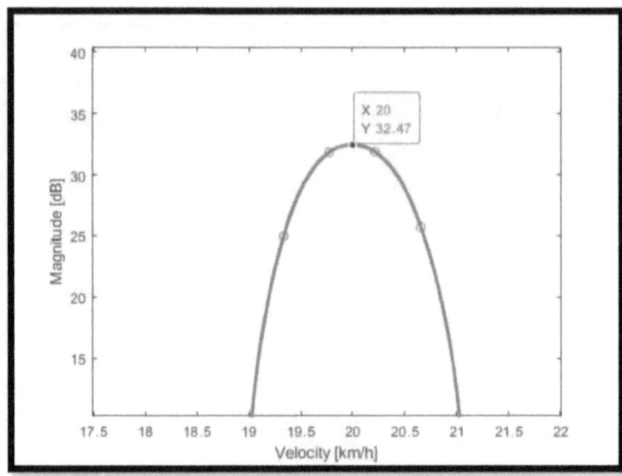

Figure 7-18: Velocity estimate using interpolation for CPI =0.0512s.

The advantage when the CPI was increased, in this instance from 0.0128 s to 0.0256 s and 0.0512 s, was an increased FFT resolution, SNR and a better velocity estimate.

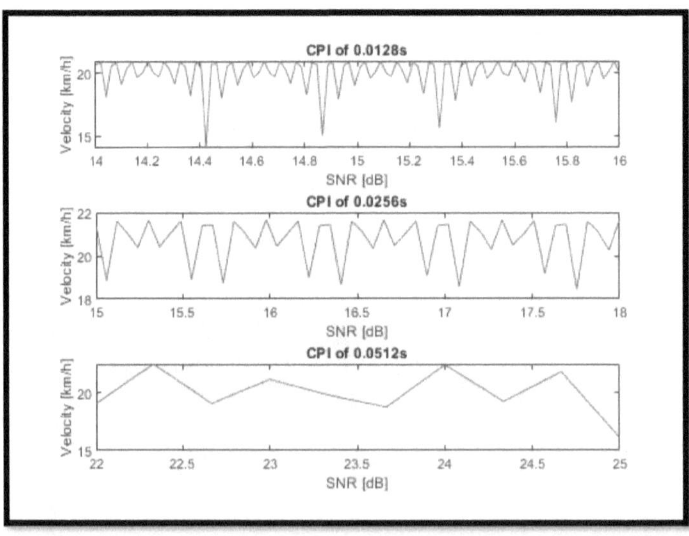

Figure 7-19: Velocity estimate as a function of SNR for vehicle travelling at 20 km/h.

It can be seen from Figure 7-19 that using the interpolation method over several Doppler profiles for each experiment shows the velocity estimate fluctuating around 20 km/h. The obvious consequence of having high CPI results in fewer fluctuations and a higher SNR. Figure 7-19 shows that for a CPI of 0.0128 s the mean velocity estimate was 20.319 km/h, for a CPI of 0.0256 s the mean velocity estimate was 20.18 km/h and for a CPI of 0.0512 s the velocity estimate was found to be 20.06 km/h. It was also observed that there was no correlation between a high signal to ratio and improved velocity estimation, the cause for a high SNR and an improved velocity estimate was an increase in CPI. The higher the CPI, the more samples to integrate over as well as a finer velocity resolution.

This experiment was repeated with the electric car travelling again at 40 km/h with the same CA-CFAR parameters of N= 50 and G= 12.

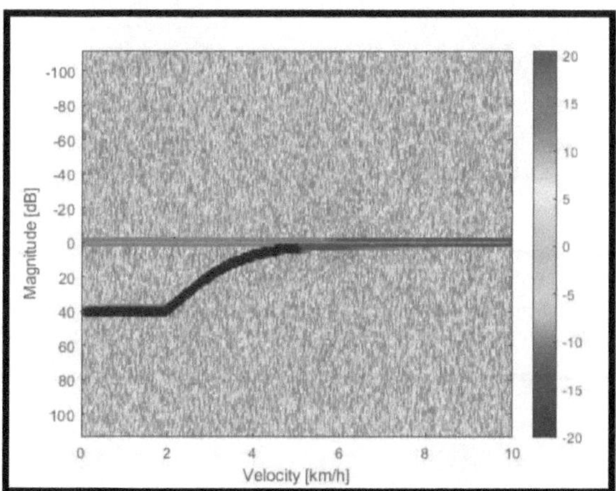

Figure 7-20: A spectrogram of electric car travelling at a constant 40km/h for the first 2 seconds then decelerating to zero in 3 seconds with a CPI of 0.0128s.

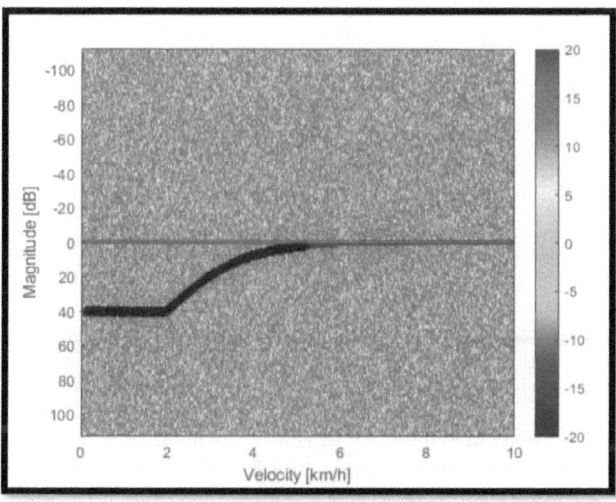

Figure 7-21: A spectrogram of electric car travelling at 40 km/h for the first 2 seconds then decelerating to zero in 3 seconds with a CPI of 0.0256 s.

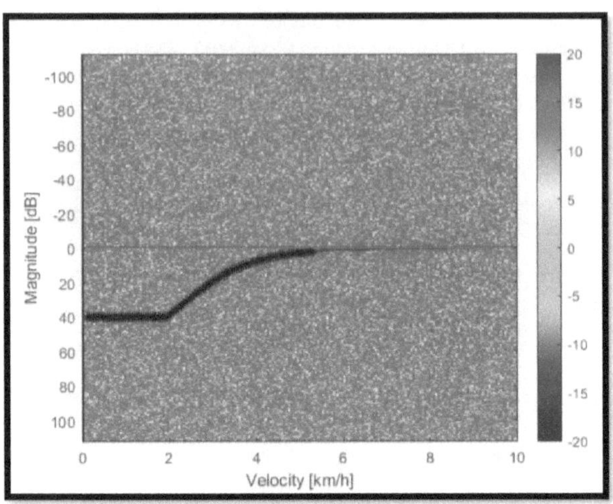

Figure 7-22: A spectrogram of electric car travelling at 40 km/h for the first 2 seconds then decelerating to 0 km/h in 3 seconds with a CPI of 0.0512 s.

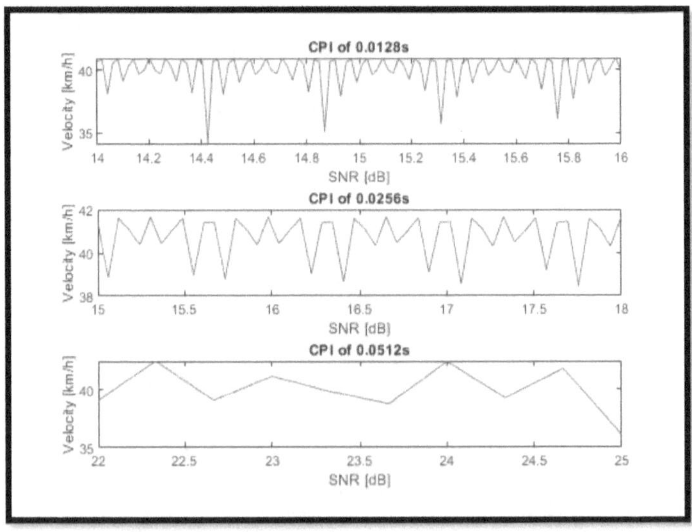

Figure 7-23: Velocity estimate as a function of SNR for vehicle travelling at 40 km/h.

Figure 7-20 proves the precision of the system, again showing a constant speed corresponding to a constant velocity line on the spectrogram. The 1st detection was made at t =0 s which corresponds to a 40 m detection range and a 100% detections rate by the CA-CFAR whilst avoiding any false alarms within the 100k samples in the spectrogram. Figure 7-21 shows that by increasing the CPI to 0.0256 s, there was an improvement in the SNR at distances of 40 m and that the 100% detection rate was maintained together with zero false detections associated with having a P_{FA} of 10e-6; this P_{FA} significantly reduces the likelihood of a false detection appearing within the 100k samples. Figure 7-22 shows that using a CPI of 0.0512 s, the detection distance of 40 m was also achieved and all the benefits of having a high SNR were maintained; this includes a 100% detection rate and zero false detections within the 100k analysed samples of the spectrogram. In Figure 7-23, It can be seen that the velocity estimate vs SNR graph, the mean velocity estimate is 40.319 km/h using a CPI of 0.0128 s, 40.18 km/h for a CPI of 0.0256 s and 40.06 km/h for a CPI of 0.0512 s was very similar to that of Figure 7-19, both graphs have the same mean velocity estimates and velocity error values. This result is from using the same interpolation method, which was the "spline" method. The SNR was decorrelated from the velocity estimate, meaning that an improved SNR did not lead to a better velocity estimate.

The number of samples used to calculate the mean velocity estimate differed when changing the CPI, but since the total number of samples in the spectrogram remained the same, there was a trade-off between precision and accuracy. The higher the CPI was chosen, the finer the velocity resolution, but the Doppler profiles decreased. Which means for a CPI of 0.0128 s there were 128 samples in the samples in the Doppler dimension but 781 Doppler profiles. This means the system had fewer samples to compute the velocity estimate, but the estimate this computation resulted in was more precise than that of computation with fewer Doppler profiles. Whist a CPI of 0.0512s had 512 samples in the Doppler dimension but only 195 Doppler profile; this means the velocity estimate computed using 512 samples was more accurate than the estimate using 128 samples in the Doppler dimension but less precise since it had fewer Doppler profiles.

The accuracy of a measurement is measured against the ground truth measurement, and the ground truth measurement is dependent on the accuracy and precision of the instrument used to measure it. Thus, an estimate can never be more accurate than the ground truth measurement even when the instrument used to make the estimate is more accurate than the instrument used to measure the ground truth measurement [36]. Since the ground truth measurement of the electric vehicle had an uncertainty of ± 0.096 km/h [68], the velocity estimate made by the radar is a combination of the uncertainty of the ground truth measurement and mean velocity measurement. The CRLB predicts the limit of the precision; this statistic indicates the best achievable precision of an estimate. CRLB is calculated using the SNR of a measurement and frequency resolution of the system.

The effective detection rate can be found by taking the vehicle observation time of the true detections that are visible on the spectrogram and dividing it by the total vehicle observation time. This result and results of the experiments made in this section were summarized in Table 7-2.

Table 7-2: Detection characteristics of the system as function of CPI.

Coherent processing time (CPI)	Minimum detection SNR [dB]	Cramer-Rao lower bound precision [km/h]	Electric Vehicle ABS uncertainty [± km/h]	Mean velocity estimate uncertainty [± km/h]	Combined velocity estimate uncertainty [±km/h]	Detection distance [m]	Detection rate [%]
0.0128 s	13.25	±0.021	±0.096	±0.319	±0.415	40	100
0.0256 s	14	±0.01	±0.096	±0.180	±0.276	40	100
0.0512 s	22.09	±0.004	±0.096	±0.060	±0.156	40	100

Table 7-2 shows that the minimum detection SNR was found to increase when increasing the CPI, Requirement 2 stated that the maximum detection must be 40 m and this was found to be the case for this system no matter which CPI that was chosen. The velocity precision was required to fall between ± 2 km/h and it was found that each CPI tested produced a velocity estimate with an uncertainty that which was lower than the one which was stipulated in the requirement, the calculated CRLB for this system was found to be ± 0.021km/h, ± 0.01 km/h and ± 0.004 km/h for a CPI of 0.0128 s,0.0256 s and 0.0512 s respectively at 40 m.

It must also be acknowledged that the uncertainty of a measurement indicates the quality of the measurement. Thus, the actual velocity uncertainty of the measurements are $\pm$ 0.415 km/h, ± 0.276 km/h and ± 0.156 km/h for a CPI of 0.0128 s, 0.0256 s, and 0.0512 s respectively. These results are only for the velocity ranges between 20 - 40 km/h.

In these experiments, the car was travelling at a fixed velocity, when finding the accuracies of these velocity measurements, the speedometer readings of the car were used as ground truth values. When performing the experiments for the car travelling at 20 km/h, the measured speed by the radar was an average 20.186 km/h this means the car achieved an accuracy of 99.07%. When the car was travelling at 40 km/h the measured speed by the radar was an average of 40.192 km/h which, means the average speed was found to be 99.52%.

Since the SNR for a CPI of 0.0512 s was higher than the required SNR for a detection at 40 m, this means that using this processing parameter the radar can obtain detections further than 40 m, but the exact distance must still be found using more empirical data. The detection rate was found to be 100% since there were no missed detections in the spectrogram.

The recommended parameters for the radar signal processing were summarized in Table 7-3.

Table 7-3: Optimal signal processing parameters

Coherent Processing Interval [s]	Number of reference cells N	Number of Guard cell G
0.0512	50	12

7.4   Experiment C1: Maximum Speed Detection

The purpose of this experiment was to show that the radar could detect vehicles travelling at speeds of up to 60 km/h; this was part of the 1st requirement in the technical specifications. In this experiment, the radar was placed at the side of the road, as shown in Figure 7-5 The driver first had to do a lap around campus to accelerate to the required speed. The recording began when the car was at 70 m away from the radar.

This experiment was performed under the following restrictions from the security detail of the CSIR to ensure the safety of non-motorized and motorized users of the CSIR roads. The vehicle must only accelerate to a speed of 60 km/h; once the speed has been reached, the vehicle must decelerate to a safer speed. Figure 7-24 is the resulting spectrogram from the experiment using the optimal signal processing parameters detailed in Table 7-3.

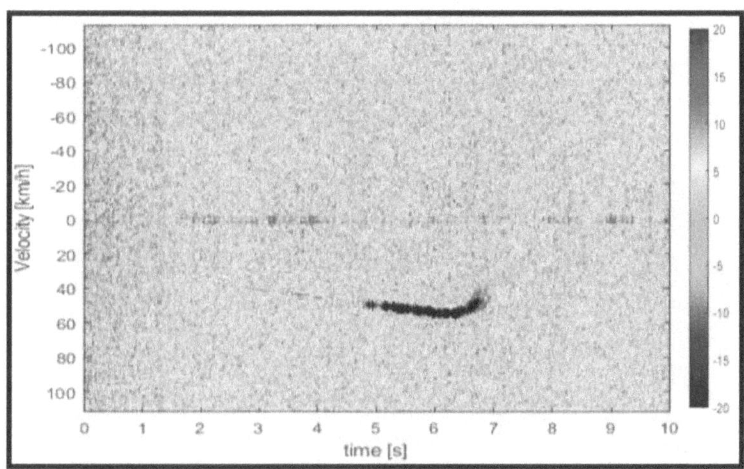

Figure 7-24: A spectrogram of a vehicle travelling at a peak of 58 km/h with CPI of 0.0512 s.

Figure 7-24 illustrates a vehicle travelling at a peak velocity estimate of 58 km/h towards the radar. This experiment showed that the radar could capture these speeds. Unfortunately, there were missed detections in the spectrum. The detection rate was calculated as the number of true detections over the number of true detections + missed detections. There was a total of 195 Doppler profiles in the spectrogram for a CPI of 0.0512 s, and visually the detections begin from the 5th second till the 7th second meaning the visual detections occur for 40 Doppler profiles starting from the 97 Doppler profile. There was a total of 39 detections meaning the observed detection rate was 97.5% for a CPI of 0.0512 s; there were no observable false detections in the spectrogram. The velocity estimate shown in Figure 7-24 was also restricted by the Doppler resolution, so interpolation must be done in order to get a more accurate estimate of the velocity. This process was illustrated in Figure 7-25.

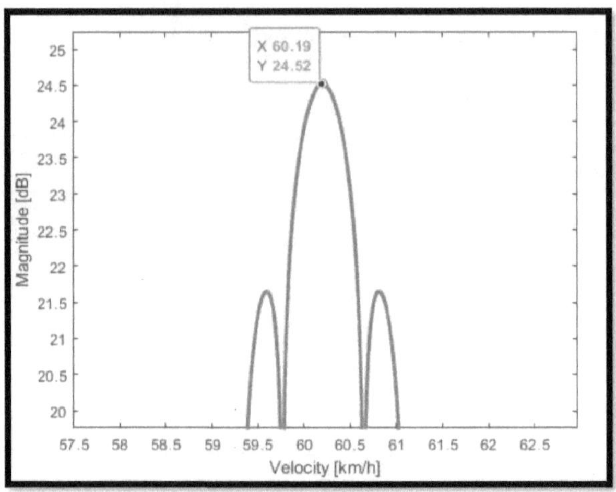

Figure 7-25: Velocity estimate of 60.19 km/h using interpolation for CPI =0.0512s.

The velocity estimate using the interpolation method yielded an estimate of 60.19 km/h, this velocity estimate has a velocity measurement uncertainty of ±0.19 km/h which was added to that of the ground truth measurement and the resultant velocity uncertainty was ±0.286 km/h. The velocity measurement uncertainty was larger because the vehicle speed was not constant, and this estimate was made using one Doppler profile. The accuracy of the measurement was found by taking the ground truth measurement of 60 km/h, according to the speedometer of the car, and finding the difference with the velocity estimate of the radar which is 60.19 km/h; the result is a 99.6% accuracy.

The next question was that of the maximum detection range when the car travelled at the average detected speed, in order for the driver to react to the instruction to slow down they require 2 s when travelling at 60 km/h. The signal exists for at least 2 seconds, and the estimated velocity was found to be 60.286 km/h or 16.746 m/s, which means the car travelled approximately 33.49 m.

It must be asked what happens when the vehicle travels at higher speeds than 60 km/h. It is known that the radar was capable of detecting speeds that are greater than 60 km/h as the radar module had maximum velocity rating of 1125 km/h [37]. It must be shown that the signal processor can detect speeds greater than 60 km/h. Therefore, a spectrogram with a simulated target travelling speed of 100 km/h was made. This process was shown in   Figure 7-26.

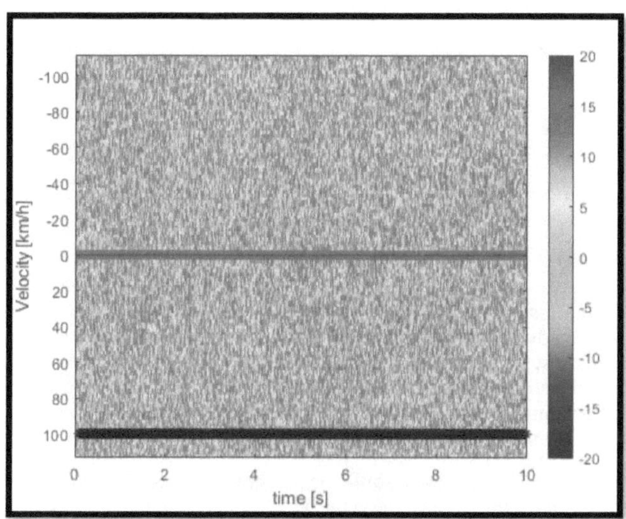

Figure 7-26:Simulated target travelling at 100km/h

It can be seen in Figure 7-26 that a target travelling at 100 km/h in simulated noise and clutter can be detected easily; this shows that the CA-CFAR is indeed capable of observing targets with speeds higher than 60 km/h.

7.5   Summary

In this chapter, the acceptance tests were conducted; these consisted of several experiments that aimed to quantify the real-world performance of the proposed radar system. The first experiment quantified the noise intensity for varying coherent processing intervals. The assumption that the environment the radar was to be placed in is homogenous was tested and proven. It was also observed that the true false alarm rate was 1.08e-6, which means the relative P_{FA} error was 8%. The second experiment exploited the homogenous nature of the system to obtain the SNR values as a function of range. It was found that SNR increased as the vehicle approached the radar. The third experiment investigated the velocity and detection accuracy of the system using different CPI and deciding what the most optimal signal processing parameters were. It was found that a CPI of 0.0512 s produced the largest SNR at 40 m and had the lowest velocity measurement uncertainty from the CPI values tested. The last experiment demonstrated that the radar could detect speeds of 60 km/h and that the detector could identify targets travelling at speeds of up to 100 km/h. The velocity accuracies reached were above 99%, which means this system was both accurate and precise.

Chapter 8

Conclusion

8.1   Conclusion

This study details the design and prototyping of a traffic calming radar solution. The radar system consisted of a radar module, DSP, power system ADC and the enclosure. This system was designed to have a comparable system performance with the systems reviewed in Section 2.2 Table 2-7 and Table 2-13. The most desirable specifications listed in these summaries include the low measurement uncertainties, which means that these systems produced very precise and accurate measurements, low latency measurements and maintenance-free operation.

The summary in Table 2-13, highlighted sophisticated power systems which feature rechargeable batteries and solar panels; this enabled long term maintenance free operation and reliability. All these features were considered desirable for the proposed system, and in Section 4.2, the proposed system requirements were formulated in accordance with the guidelines set up by the user requirements.

The proposed system was designed using the theoretical foundations laid out in Chapter 3; this prompted the adoption of the CW radar architecture over pulse-Doppler and FMCW architectures. The simplicity and reliability of CW Doppler processing was the main contributor to this choice. The design parameters were also refined using principles featured in this chapter.

In Chapter 4, the actual system requirements were stated; the success of the study was predicated on the successful fulfilment of the system requirements outlined in Section 4.4. This study showed that a system comparable to commercial systems was feasible at below cost, by the successful completion of the application test procedures summarized in Section 4.4 Table 4-2 and the bill of the materials found in Section 5.9 Table 5-16.

Chapter 5 saw the development of the system design using the parameters outlined in the previous chapters. The radar module architecture was discussed in detail in Section 5.2.1, and the specifications of the radar module presented in Section 5.2.2; this led to a comparison of available radar modules that met the minimum specifications in Table 5-1. A suitable radar module was chosen in the IPS-154 radar module; it presented a theoretical SNR of 50.72 dB which was considered suitable for this application. The succeeding sections in this chapter saw the comparison of suitable ADC and DSPs in the efforts of obtaining devices that can deliver high-quality, low latency velocity estimates. The chosen ADC and DSP allowed for a theoretical velocity estimation computation time of 13 ms, 26 ms and 52 ms for a CPI of 0.0128 s, 0.0256 s and 0.0512 s, respectively as shown in Table 5-9. The power system, display module and enclosure were specified, and the total cost of the proposed system was calculated. Since the proposed system is costly and the construction of this system is out of the scope of this study; The chapter concluded with the specification of the data-collection module which would enable experiments that could prove the validity of the proposed system.

Chapter 6 saw the integration and testing of the data-collection system and signal processing algorithms; the various issues that plagued the system included the frequency harmonics resulting from the inconsistent grounding of the power supply unit of the data collection system. Another issue encountered was the discontinuous lines in the frequency spectra resulting in disconnection of the radar module from the ADC. These issues were addressed, and the solutions thereof allowed for the experiments detailed in Section 4.4 to be carried out with no technical difficulties.

The tests detailed in Section 4.4, were able to show that with sufficient SNR, the system would be able to make reliable detections of small vehicles at a distance of 40 m. This was shown by both experiment A2 and B1 in Section 7.2 and 7.3, respectively. The radar speed calming system was designed to have a detection probability of 90% and a probability of false alarm of 1e-6. The actual system achieved detection rates of 100% for all CPIs tested. The effective false alarm was found to be 1.08e-6 which had a relative error of 8%.

It was also proven using the ATPs that a velocity accuracy above 99% at 40 m was achievable using affordable radar hardware and basic signal processing for cars travelling between 20 to 60 km/h. Experiment B1 in Section 7.3 illustrated that when using an electronic vehicle to obtain ground truth velocities. The radar was able to produce velocity estimates comparable to the ground truth velocities with a velocity measurement uncertainty of ±0.415 km/h, ±0.276 km/h, ±0.156 km/h using an of CPI 0.0128 s, 0.256 s and 0.0512 s respectively when tested using an electric vehicle travelling at a constant speed.

Experiment C1 in Section 7.4 roved that the proposed radar-based traffic calming solution was able to detect velocities of up to 60 km/h, which is crucial to enforcing campus speed calming. It was also shown that the detection algorithm could detect targets travelling at speeds of up to 100 km/h.

The pole design is also able to handle a total weight of 16618 kg, as shown in Section 5.9. The proposed display is able to reach an illumination that is observable from 40 m away in day and night conditions. The enclosure that was designed could be tailor-made at a reasonable cost, and the entire system could be built at the cost of R12423. The total cost of the components that made up the system was R10848, and the labour to build this radar was quoted at R1575.

Table 8-1: Summary of conclusion

Requirement	Status
1	Satisfied
2	Satisfied
3	Satisfied
4	Satisfied
5	Satisfied
6	Satisfied
7	Satisfied

References

[1] I. M. Lockwood, "ITE Traffic Calming Definition," *ITE Journal,* vol. 67, July 1997, pp. 22-24.

[2] M. S. Rajani, "Smart speed bumps," M.S. thhesis,Dept. Civil. Eng., Univ.Texas, Arlington, Tex, 2015.

[3] G. Hoole and C. Gibson, The New Official K53 manual, Johannesburg: Penguin Random House, 2013.

[4] K. Chang, M. Nolan and N. L. Nihan, "Radar speed signs on neighborhood streets: an effective traffic calming device?," ITE annual meeting and exhibit, Lake Buena Vista, FL, 2004.

[5] Wikipedia. (2010, October 1). *Radar speed sign,* [Online]. Available: https://en.wikipedia.org/wiki/Radar_speed_sign. [Accessed 23 February 2019].

[6] L. E. Y. Mimbela and L. A. Klein. (2000, September 23). *U.S. Department of Transportation Federal Highway Administratioin,* [Online]. Available: https://www.fhwa.dot.gov/ohim/tvtw/vdstits.pdf. [Accessed 15 July 2019].

[7] T. Litman, "Traffic Calming Benefits, Costs and Equity Impacts," Victoria Transport Policy Institute, Victoria, 1999.

[8] A. Riid, J. Kaugerand, J. Ehala, M. Jaanus and J.-S. Preden, "Application of a Low-Cost Microwave Radar toTraffic Monitoring," in *16th Biennial Baltic Electronics Conference,* Tallinn (BEC), 2018, pp.1-4.

[9] C. Wolff. (2003, January 9). *Radar basics,* [Online]. Available: http://www.radartutorial.eu/08.transmitters/Waveform-Generator.en.html. [Accessed 5 May 2019].

[10] T. Butterfield. (2015, March 31). *Building Radar Speed Camera and and traffic logger with a Raspberry Pi,* [Online]. Available: http://blog.durablescope.com/tags/internet-of-things/polestar. [Accessed 10 May 2019].

[11] Cmedia. (2010, January 10). *CM6206-LX,* [Online]. Available: https://www.cmedia.com.tw/products/USB20_FULL_SPEED/CM6206-LX. [Accessed 25 May 2019].

[12] M. Longstaff-Tyrrell. (2015, September 28), *Creating a Radar Speed Detector with a STM32L476 Discovery board,* [Online]. Available: https://www.rs-online.com/designspark/creating-a-radar-speed-detector-with-a-stm32l476-discovery-board. [Accessed 18 March 2019].

REFERENCES

[13] J. Fang, H. Meng, H. Zhang and X. Wang, "Low-cost Vehicle Detection and Classification System based on Unmodulated Continuous-wave Radar," in *IEEE Intelligent Transportation Systems Conference*, Seattle, WA, 2007, pp 715-720.

[14] S.L. Anteral. (2018, January 1). *uRad*, [Online]. Available: www.urad.es. [Accessed 23 March 2019].

[15] F. Alimenti, F. Placentino, A. Battistini, G. Tasselli, W. Bernardini, P. Mezzanotte, D. Rascio, V. Palazzar, S. Leone, A. Scarponi1, N. Porzi, M. Comez and L. Roselli, "A low-cost 24GHz Doppler radar sensor for traffic monitoring implemented in standard discrete-component technology," in *2007 European Microwave Conference*, Munich, 2007.,pp. 1441 -1444.

[16] J. M. Munoz-Ferreras, J. Calvo-Gallego and F. Perez-Martinez, "Monitoring road traffic with a high resolution LFMCW radar," in 2008 *IEEE Radar Conference*, Rome, 2008, pp. 1-5. doi: 10.1109/RADAR.2008.4720895.

[17] D. Felguera-Martin , J.T. Gonzalez-Partida and P. Almorox-Gonzalez , "Vehicular Traffic Surveillance and Road Lane Detection Using Radar Interferometry," *IEEE Transections on Vehicular Technology,* vol. 61, no. 3, pp. 959-969, March 2012. doi: 10.1109/TVT.2012.2186323.

[18] Huston Radar. (2009, January 19). *PNL10 OEM radar speed sign kit,* [Online]. Available: https://houston-radar.com/pnl10-oem-radar-speed-sign.php#.XQllshYzb3g. [Accessed 6 March 2019].

[19] Wanco Inc. (2019, February 18). *Pole Mount Speed Signs,* [Online]. Available: https://www.wanco.com/wp-content/uploads/2018/12/specs_SpeedSignsPolemount_WSDP.pdf. [Accessed 20 March 2019].

[20] Trafficlogix. (2016, October 23). *11 inch Radar signs,* [Online]. Available: https://trafficlogix.com/safepace-100/. [Accessed 8 May 2019].

[21] Monitor Systems. (2019, February 17). *MSPM-2 numeric-radar-speed-sign-2-or-3-digit-display-for-kmh,* [Online]. Available: https://monitorsystems.net/products/pole-mounted-speed-signs/. [Accessed 9 May 2019].

[22] Radarsign. (2009, August 3). *TC-400 Portable Radar Speed Sign*, [Online]. Available: https://www.radarsign.com/radar-speed-signs/tc-400-portable-radar-speed-sign/. [Accessed 1 May 2019].

[23] MPH Industries. (2014, March 6). *Speed Monitor F advisory sign*, [Online]. Available: https://www.mphindustries.com/speed-monitor-f/. [Accessed 6 May 2019].

[24] ICASA, *National radio frequency plan 2018 for 8.3 kHz – 3000 GHz*, Government Printing Works, Pretoria, 2018.

[25] F. Maasdorp, EEE5120Z, Lecture slides,Topic: "Introduction to Electronic Defence", Department of Electrical Engineering, Univ. Cape Town , Cape Town, May 11, 2018.

[26] M. A. Richards, J. A. Scheer and W. A. Holm, Principles of Modern Radar: *Basic Principles*, Raleigh: SciTech Publishing, 2010.

REFERENCES

[27]   M. Martorella, EE5119Z, Lecture slides, Topic: "Introduction to Radar," Department of Electrical Engineering, University of Cape Town, Cape Town, Feb. 17 ,2017.

[28]   G. L. Charvat. (2014, February 2). *Guest post: Try radar for your next project*, [Online]. Available: https://hackaday.com/2014/02/24/guest-post-try-radar-for-your-next-project/. [Accessed 16 March 2019].

[29]   W. Weidmann, *Application Note III.* (1999) InnoSenT GmbH, Donnersdorf.

[30]   T. Schipper, J. Fortuny-Guasch , D. Tarchi, L. Reichardt and . T. Zwick, "RCS Measurement Results for Automotive Related Objects at 23-27 GHz," IEEE Xplore, Karlsruhe, 2011.

[31]   J. Kowalski, "Receiver sensitivity/noise," 15 December 2006. [Online]. Available: https://www.phys.hawaii.edu/~anita/new/papers/militaryHandbook/rcvr_sen.pdf. [Accessed 29 May 2019].

[32]   N. Wireless. (2018, February 16). *From Analog to Digital – Part 2: The Conversion Process*, [Online]. Available: https://www.nutaq.com/blog/analog-digital-%E2%80%93-part-2-conversion-process. [Accessed 1 April 2019].

[33]   R. E. Ziemer, W. H. Tranter and D. R. Fannin, "Analog -to-Digital Conversion," in *Signals and System: Continuous and Discrete*, ,4th ed. Natick, Pearson Prentice-Hall, 1998, p. 366.

[34]   S. M. Kay, Fundimentals of Statistical Signal Processing,Vol.1: Estimation Theory, Upper Saddle River: Prentice hall, 1993. pp.157-173.

[35]   Y. PU, F. Chen, X. Pan, J. Liang, H. Peng and Y. Wu, "The legibility of LED traffic guide signs in urban tunnels," in *4th International Conference on Transportation Information and Safety (ICTIS)*, Banff, 2017 pp. 1111-1116. doi: 10.1109/ICTIS.2017.8047909.

[36]   Deep Cycle Marine Battery. (2010, April 25). *Best 100 Ah AGM Deep Cycle Marine Battery*, [Online]. Available: DeepCycleMarineBattery.com. [Accessed 2019 August 20].

[37]   Innosent GmbH, "IPS-154," "IPS-154," IPS-154 Datasheet, Feb. 2019 [Revised Oct. 2015].

[38]   H. R. inc. (2015, Januaray 1). *PD300/PD310 OEM FMCW*, [Online]. Available: http://houston-radar.com/PD300-OEM-doppler-speed-sensor.php#.XIFMBiIzb3h. [Accessed 27 February 2019].

[39]   Innosent GmBH, "Radarsensor IPS-355", IPS-355 datasheet, Feb. 2019 [Revised 1 Jan. 2013].

[40]   Innosent GmBH, "Radarsensor IPS-154", IPS-355 datasheet, Feb. 2019 [Revised 1 Jan. 2013].

[41]   Innosent GmBH, "Radarsensor IPS-937," IPS-355 datasheet, Feb. 2019 [Revised 1 Jan. 2013].

[42]   Innosent GmBH, "Radarsenor IPS-280," IPS-355 datasheet, Feb. 2019 [Revised 1 Jan. 2013].

REFERENCES

[43] Innosent GmBH, "Radarsensor IPS-144," IPS-355 datasheet, Feb. 2019 [Revised 1 Jan. 2013].

[44] Analog Devices. (2003, January 16). *A/D Convertors*, [Online]. Available: www.analog.com. [Accessed 3 October 2020].

[45] P. T. Ltd. (2015, January 1). *PicoScope6UsersGuide*, [Online]. Available: https://www.picotech.com/download/manuals/PicoScope6UsersGuide.pdf. [Accessed 11 February 2019].

[46] Maxim Intergrated. (2020, August 16). *Single-Supply, Low-Power, 2-Channel, Serial 8-Bit ADCs*, [Online]. Available: https://www.maximintegrated.com/. [Accessed 3 October 2020].

[47] Silicon Labs. (2020, July 17). *C8051F2xx Small Form Factor Microcontrollers*, [Online]. Available: https://www.silabs.com/. [Accessed 3 October 2020].

[48] STMicroelectronics.(2019, June 30). *Development Boards, Kits, Programmers*, [Online]. Available: https://www.st.com/. [Accessed 3 October 2020].

[49] H. Rohling, "Ordered statistic CFAR technique - an overview," in *201112th International Radar Symposium (IRS)*, Leipzig, 2011, pp. 631-638.

[50] P. P. Gandhi and C. S. Witte, "Performance of distributed CFAR processors in nonhomogeneous background," in *Proceedings of 1994 IEEE National Radar Conference*, Atlanta, GA, USA, 1994. pp. 212-217. doi: 10.1109/NRC.1994.328126.

[51] P. P. Gandhi and S. A. Kassam, "Optimality of the cell averaging CFAR detector," *IEEE Transactions on Information Theory*, vol. 40, no. 4, pp. 1226-1228, July 1994. doi: 10.1109/18.335950.

[52] A. A. Melebari and M. Y. Abdul Gaffar ,EEE5105Z, Lecture code, Topic: "CA_CFAR Simulator.m", Department of Electrical Engineering, University of Cape Town, Cape Town May. 26. 2017

[53] N. Longqiang ,. G. Shesheng and X. Li, "Improved Probabilistic Data Association and Its," in *2011 International Conference on Electronics, Communications and Control (ICECC)*, Ningbo, 2011, pp. 293-296. doi: 10.1109/ICECC.2011.6066629.

[54] The Pyzo team. (2015, March 16). *Python vs Matlab*, [Online]. Available: https://pyzo.org/python_vs_matlab.html. [Accessed 17 Spetember 2019].

[55] The julia project. (2019, August 21). *Julia 1.2 Documentation*, [Online]. Available: https://docs.julialang.org/en/v1/. [Accessed 17 September 2019].

[56] P. M. Embree and D. Danieli, "C++ Libraries," in *C++ Algorithms for Digital Signal Processing*, New Jersey, Prentice Hall, 1998, pp. 360-361.

[57] Parewa Labs Pvt. Ltd (2011, June 30). *Learn Java Programming*, [Online]. Available: https://www.programiz.com/java-programming. [Accessed 16 October 28].

[58] M. Reznicek and P. Bezousek, "Commercial CW Doppler radar design and application,," in *27th International Conference Radioelektronika (RADIOELEKTRONIKA)*, Brno, 2017, pp. 1-5. doi: 10.1109/RADIOELEK.2017.7937577.

[59] A. J. Maren (2014, February 21). *GPUs, CPUs, MIPs, And Brain-Based Computation*, [Online]. Available: http://www.aliannajmaren.com/2014/02/21/gpus-cpus-mips-and-brain-based-computation/. [Accessed 31 May 2019].

[60] E. Upton. (2014, January 30). *Accelerating Fourier transforms using the GPU*, [Online]. Available: https://www.raspberrypi.org/blog/accelerating-fourier-transforms-using-the-gpu/. [Accessed 31 May 2019].

[61] E. F. Schubert. (2006, January 1). *Human eye sensitivity and photometric quantities,* [Online]. Available: https://www.ecse.rpi.edu/~schubert/Light-Emitting-Diodes-dot-org/Sample-Chapter.pdf. [Accessed 31 May 2019].

[62] Varta-automotive. (2020, May 15). *AGM Batteries vs. Gel Batteries*, Clarios, [Online]. Available: https://www.varta-automotive.com/en-nz/varta-battery-support/battery-basics/agm-batteries-vs-gel-batteries. [Accessed 9 Octorber 2020].

[63] Sell-SA. (2020, July 17). *12 V 50 Ah deep-cycle gel solar battery manufactured by Gamistar*, [Online]. Available: https://www.sellsa.co.za/50ah-12v-deep-cycle-gel-lead-acid-vrla-solar-battery-gamistar. [Accessed 9 October 2020].

[64] D. Llorens. (2012, May 5). *Do solar panels work in cloudy weather?*, [Online]. Available: https://www.solarpowerrocks.com/solar-basics/how-do-solar-panels-work-in-cloudy-weather/. [Accessed 09 October 2020].

[65] R. C. Hibbeler, Machanics of Materials, Jurong: Pearson Hall, 2011.

[66] J. Mahachi, Design of Structural Steelwork to SANS 10162, Randburg: Xsi-tek, 2013.

[67] Advanced Navagation. (2017, September 18). *Spatial Dual Reference Manual*, [Online]. Available:
https://www.advancednavigation.com/sites/default/files/product_documents/Spatial%20Dual%20Reference%20Manual.pdf. [Accessed 16 August 2019].

[68] Apec Tech Mate, "Technical Guide: ABS Sensors," TECHMATE, London, England, Tech. Report, 24 Apr. 2017.